Background Math
for the Board of Certified Safety Professionals'
Safety Certification Examinations

Glenn Young, CSP

First Edition

Background Math for the Board of Certified Safety Professionals' Safety Certification Examinations © 2003 American Society of Safety Professionals

Published by the American Society of Safety Professionals, Park Ridge, IL.

No part of this publication may be reproduced, stored in a retrieval system, or transmitted in any form or by any means, electronic, mechanical, photocopying, recording, scanning, or otherwise, except as permitted under Section 107 or 108 of the 1976 United States Copyright Act, without prior written permission of the Publisher.

Limit of Liability/Disclaimer of Warranty: While the publisher and author have used their best efforts in preparing this book, they make no representations or warranties with respect to the accuracy or completeness of the contents of this book and specifically disclaim any implied warranties of merchantability or fitness for a particular purpose.

Managing Editor: Michael Burditt
Copyediting, Text Design, and Composition: Sue Knopf, Graffolio
Cover Design: Michael Burditt and Tina Angley, ASSP

ISBN: 978-1-885581-45-7

Printed in the United States of America

23 22 21 20 19 18   8 9 10 11 12 13 14 15

Foreword

Math is an important part of everyday life and an integral part of the skills necessary to become certified in the safety profession. Many who pursue certification have long since completed their college math courses and have not actively used the math skills they once had. This book, which is written in a straightforward and easy-to-use format, provides the basics necessary to successfully negotiate the math included on the certification exams. It can also serve as a handy primer for those that already have their credentials.

C. David Langlois, CSP, and Donald S. Jones, P.E., CSP

CONTENTS

CHAPTER ONE: CALCULATOR SELECTION AND USE 1
- B.C.S.P. Rules for Calculators 1
- Calculator Hierarchy of Operations 3
- Calculator Strategies for Examinations 4

CHAPTER TWO: FRACTIONS, ETC. 5
- Representing Fractions as Decimals or Percents 5
- Multiplying Fractions 8
- Dividing Fractions 11
- Adding and Subtracting Fractions 16
- Reciprocals 21
- Proportions 22
- Rounding 24
- Absolute Value 27

CHAPTER THREE: EXPONENTS, ROOTS, AND LOGS 31
- What Is an Exponent? 31
- Rules of Exponents 32
- What Is a Root? 36
- Logarithms and Antilogs 39

CHAPTER FOUR: SYSTEMS OF MEASUREMENT 45
- English Units 45
- Metric Units 48
- Other Measurement Systems 52
- Conversions 55
- Dimensional Analysis 58

CHAPTER FIVE: SCIENTIFIC AND ENGINEERING NOTATION 67
- Scientific Notation 67
- Engineering Notation 71

CHAPTER SIX: ALGEBRAIC PROPERTIES AND SIMPLE EQUATIONS........ 75

Variables 75
Commutative Properties 75
Associative Properties 76
Distributive Property 77
Factoring 78
Collecting Like Terms 78
Multiplying Polynomials 78
Order of Operations 81
Rules of Equations 82
Solving Equations with a Variable on One Side 84
Solving Equations with Variables on Both Sides 86
Solving Equations with Multiple Variables 87
Solving Simultaneous Equations 88
Proportions 92

CHAPTER SEVEN: APPLIED ALGEBRA.. 103

Sets, Subsets, and Venn Diagrams 103
Set Symbols 104
Mixtures 106
Graphing, Slopes, and Intercepts 108
Zero Product Property 112
Quadratic Equations 112
Word Problems 114

CHAPTER EIGHT: GEOMETRY.. 131

CHAPTER NINE: TRIGONOMETRY 145

Special Properties of Right Triangles 146
Trigonometric Functions 146
Law of Cosines 150
Law of Sines 150
Scaffold Problems 154
Sling Problems 156
Friction 161
Ramp Problems 162
Multiple Triangles 163

CHAPTER TEN: BOOLEAN ALGEBRA 179

Fault Tree Analysis 18

Abbreviations and Symbols Used in This Book 18
Index 19

TABLES

Table 1: Calculator Requirements 2
Table 2: Calculator Preferences 2
Table 3: English Measurement Liquid Conversions 46
Table 4: Time Conversions 46
Table 5: English Measurement Distance Conversions 47
Table 6: Metric Prefixes 50
Table 7: Alternate Measurement System Units 52
Table 8: SI Unit Conventions 53
Table 9: Conversions Given by *Candidates' Handbook* 55
Table 10: Conversion Factor Magic Numbers 57
Table 11: Set Symbols 104
Table 12: Two-Dimensional Shapes 132
Table 13: Three-Dimensional Shapes 133
Table 14: Special Properties of Right Triangles 146
Table 15: Trig Functions 146
Table 16: Boolean Postulates 180
Table 17: Common Fault Tree Symbols 181
Table 18: Logic Gates 182
Table 19: Abbreviations and Symbols Used 189

CHAPTER ONE

Calculator Selection and Use

> **Application:** Required for ASP, CSP, OHST, and CHST examinations.
>
> **Discussion:** Scientific calculators are a necessary tool for exam candidates. The Board of safety Professionals (BCSP) limits "acceptable" calculators to those listed in their *Candidates' Handbook*. Several brands are available and acceptable. This chapter is intended to guide you through criteria for selection and strategies for use.

BCSP Rules for Calculators

The BCSP allows some latitude in selecting your calculator, and you may bring two calculators (but no more) to the exam. BCSP's *Candidates' Handbook* provides a list of specific brands and models that are acceptable. If you have a calculator that is not on the approved list when you go for your examination, you won't be allowed to take your test.

These rules may change. Get the latest information by visiting the BCSP's website (for ASP & CSP examinations) at www.bcsp.org. Information on calculator restrictions for the Occupational Health & Safety Technologist (OHST) and Construction Health & Safety Technologist (CHST) examinations can be found at the website of the Committee for Certification of Health, Environmental, and Safety Technologists at www.cchest.org.

Although any calculator that meets the above limitations can be used, there are certain functions that a calculator *must* provide in order for you to complete the tests on time. Required minimum functions include:

Logarithm and inverse log (10^x) functions
Natural logarithm (LN) and inverse natural log (e^x) functions
Scientific notation
A universal power key (y^x)
A universal root key ($\sqrt[x]{y}$)

CHAPTER 1

A "pi" key (π)
Trig functions including *SIN, COS, TAN*
Inverse trig functions including SIN^{-1}, COS^{-1}, and TAN^{-1}
A factorial key (x!)
Permutation and Combination keys (nPr, and nCr)
A statistics mode with one and two variable statistics
Standard deviation keys (σx_n and σx_{n-1})
Mean key ($\bar{x}$)
Correlation key (COR)

Table 1: Calculator Requirements

If your calculator lacks any of these functions, replace it with another model. In addition to these minimum functions, it is preferable to have a calculator that also offers as many as possible of these additional functions:

Conversion from English to Metric units (cm-in, l-gal, kg-lb, $°C-°F$) and back
Reciprocal key ($1/x$)
Linear regression keys (x', y')
Pre-programmed quadratic equation
Simultaneous equations

Table 2: Calculator Preferences

Select your calculator from those on the approved list using the above criteria. Of course, the best calculator in the world is useless unless you can use it. Read your calculator's manual and practice the functions so that you understand them before you take the test.

Calculator Hierarchy of Operations

Calculators have a predetermined Hierarchy of Operations programmed in. This can be summarized as:

PEMDAS (Or "Pre Exam Math Daunts Any Student.")

First: *P*arentheses

Second: *E*xponents

Third: *M*ultiplication and *D*ivision

Fourth: *A*ddition and *S*ubtraction

This mnemonic is a reminder that operations with parentheses are calculated first; then exponents are calculated; then multiplication and division are done (starting on the left); and finally, addition and subtraction are performed (again, starting on the left).

EXAMPLE

$4 + 2 \bullet 3$ is not the same as $(4 + 2) \bullet 3$

In the first case, the Hierarchy of Operations requires that the multiplication (two times three) be performed first. This yields $4 + 6 = 10$.

Contrast this with the second case, where the parentheses require that the addition (four plus two) be performed first. This yields $6 \bullet 3 = 18$.

If a problem already happens to be in the correct order, you may be able to enter it in your calculator as it is written and get the right answer.

EXAMPLE

$\frac{1}{5} + \frac{4}{6}$ can be keyed directly into most calculators in the order it is written without using grouping parentheses. Keying *one divided by five plus four divided by six* into most calculators yields 0.8667. This works because the hierarchy of operations built into the calculator forces the second division (four divided by six) to be calculated before the addition is completed.

3

If you're not sure how your calculator handles the order of operations, always use parentheses to force the sequence you desire. Logs and roots are treated the same as exponents in the hierarchy. Within parentheses, the same hierarchy of operations applies. Brackets are sometimes used inside parentheses to group functions further.

If these instructions seem obtuse, don't worry. They will be reviewed and implemented repeatedly later in this book.

Calculator Strategies for Examinations

- Go to the exam with two calculators of the same make and model so that if you drop or break one or the battery goes dead, the other will be available. If a spare calculator is not available, consider at least bringing a spare battery.

- Most answers provided on the examinations are rounded to two or three significant digits and may not be the exact answer shown on your calculator. Select the answer closest to the exact result from the calculator.

- In order to avoid rounding errors, don't round numbers until you have achieved a final answer.

- If you can't determine the answer, guess. On these tests, there is no penalty for guessing wrong, and on any multiple-choice test, the probability of selecting the correct answer at random is one divided by the number of possible answers. Of course, an educated guess is better than a random choice. Use what you know.

- Don't waste time on problems you can't work. If you can't work a problem, take a guess, mark it for further attention, and move on. If there's time at the end of the test, go back to the problems you've marked. Remember, every question is worth only one point.

- Sometimes it is quicker to set up a problem as an equation and try each of the answers provided to see which one is correct. This is particularly true for quadratic equations if the calculator lacks a pre-programmed quadratic solver.

CHAPTER TWO

Fractions, Etc.

> **Assumptions:** You know how to add, subtract, multiply, and divide. You have a scientific calculator and basic knowledge of how to use it.
>
> **Application:** Required for ASP, CSP, OHST, and CHST examinations
>
> **Discussion:** Understanding of fractions is essential to practical mathematics. The ability to manipulate fractions effectively is a skill required constantly in algebra. Because of the importance of fractions, this section is offered as a review. If you feel that your comprehension of fractions is good, go to the end of this section and try the challenge exam. If you can work all of the problems correctly, you may skip this chapter.

Fractions are most commonly expressed as a number on top (the numerator) divided by another number on the bottom (the denominator). Examples of common fractions would be: $\frac{1}{2}$, $\frac{3}{4}$, and $\frac{75}{16}$. These are the same as 1÷2, 3/4 and 75 divided by 16, respectively. Fractions can contain variables as well as numbers. Variables are letters (lowercase letters, in this book) used to represent unknown numbers. Variables can exist in the numerator, the denominator, or both.

Representing Fractions as Decimals or Percents

Let the calculator convert all-numeric fractions to decimal equivalents.

EXAMPLE

Twelve sixty-ninths ($\frac{12}{69}$) converts via your calculator to 0.173913, or approximately 0.174. Although the button sequence varies from calculator to calculator, usually you enter the numerator, the division symbol (÷ or /), the denominator, and then the equals button.

CHAPTER 2

Once a fraction is in decimal form, convert it to a percent by multiplying by 100.

EXAMPLE

0.173913 • 100 becomes 17.3913%, or approximately 17.4%.

Conversely:

Convert a percent to a decimal equivalent by dividing by 100.

EXAMPLE

17.3913% ÷ 100 becomes 0.173913, or approximately 0.174.

In the previous examples, note that the value of the number does not change when it is converted from fraction to decimal to percent and back. These are merely different ways of expressing the same number.

EXAMPLE

$\frac{1}{2}$ is the same as 0.5, which is the same as 50%. All are equally valid ways of expressing the same number. Safety Certification exams will expect you to be able to convert from one form to another.

CONVERSION QUESTIONS

a. Convert $\frac{1}{8}$ to decimal and percent.

b. Convert $\frac{7}{16}$ to decimal and percent.

c. Convert $\frac{2.7}{256}$ to decimal and percent.

d. Convert 2.8% to decimal.

FRACTIONS, ETC.

e. Convert $\dfrac{256}{256}$ to decimal and percent.

f. Convert $\dfrac{403{,}986.75}{538{,}649}$ to decimal and percent.

g. Convert 0.028% to decimal and fraction.

h. Convert $\dfrac{1}{0.75}$ to decimal and percent.

CONVERSION ANSWERS

a. 0.125 (decimal) times 100 = 12.5%

b. 0.4375 (decimal) times 100 = 43.75%

c. 0.0105468 (decimal) times 100 = 1.05468 or approximately 1.05%

d. 2.8 ÷ 100 = 0.028 (decimal)

e. 1 (decimal) times 100 = 100%

f. 0.75 (decimal) times 100 = 75%

g. $\dfrac{0.028}{100}$ (fraction form) = 0.00028 (decimal form)

h. Approximately 1.33 (decimal) times 100 = approximately 133%

CHAPTER 2

Multiplying Fractions

Multiplication of fractions is addressed first because it is the simplest function to perform.

To multiply any two fractions, multiply the numerators and then multiply the denominators.

> **EXAMPLE**
>
> $\frac{2}{3} \cdot \frac{3}{4}$ is the same as $\frac{2 \cdot 3}{3 \cdot 4}$. This becomes $\frac{6}{12}$, which simplifies to $\frac{1}{2}$ or 0.5.

You could find the decimal equivalent by just using the calculator, but it's important to know how multiplication is performed, because when multiplication problems include variables, you won't be able to use the calculator.

> **EXAMPLE**
>
> $\frac{2}{x} \cdot \frac{y}{3}$ can't be multiplied using a calculator unless the values of the variables are known. The method of multiplying the fractions is the same, though: Numerator times numerator and denominator times denominator: $\frac{2 \cdot y}{x \cdot 3}$.

Numbers are usually placed before variables, and multiplication symbols are usually assumed (and not shown) in number-variable combinations.

> **EXAMPLE**
>
> $\frac{2 \cdot y}{x \cdot 3}$ would commonly be expressed as $\frac{2y}{3x}$.

Mixed expressions (those containing a number and a variable) in the numerator or denominator can be multiplied in the same style as above.

EXAMPLE

$\dfrac{3x}{7} \cdot \dfrac{4}{5}$ is the same as $\dfrac{3 \cdot x \cdot 4}{7 \cdot 5}$. This simplifies to $\dfrac{(3 \cdot 4) \cdot x}{35}$, and finally to $\dfrac{12x}{35}$.

The number that originally multiplied the variable (3) multiplies the number in the numerator of the second fraction (4). This also works with multiple variables.

EXAMPLE

$\dfrac{3x}{7} \cdot \dfrac{2y}{5}$ is the same as $\dfrac{3 \cdot 2 \cdot x \cdot y}{7 \cdot 5}$, which simplifies to $\dfrac{6xy}{35}$.

The technique also works when multiplying more than two fractions.

EXAMPLE

$\dfrac{1}{2} \cdot \dfrac{2}{3} \cdot \dfrac{3}{4}$ is the same as $\dfrac{1 \cdot 2 \cdot 3}{2 \cdot 3 \cdot 4}$ which simplifies to $\dfrac{6}{24}$ or $\dfrac{1}{4}$.

To multiply a whole number by any fraction, convert the whole number to a fraction by dividing it by one.

Then multiply the numerators together and the denominators together.

EXAMPLE

$4 \cdot \dfrac{1}{3}$ is the same as $\dfrac{4}{1} \cdot \dfrac{1}{3}$, which simplifies to $\dfrac{4 \cdot 1}{1 \cdot 3}$ or $\dfrac{4}{3}$.

This procedure also works if variables are present.

EXAMPLE

$x \cdot \dfrac{3}{y}$ is the same as $\dfrac{x}{1} \cdot \dfrac{3}{y}$, which simplifies to $\dfrac{x \cdot 3}{1 \cdot y}$ or $\dfrac{3x}{y}$.

CHAPTER 2

MULTIPLYING FRACTIONS QUESTIONS

a. $\dfrac{3}{4} \cdot \dfrac{7}{13}$

b. $\left(\dfrac{7}{8}\right)\left(\dfrac{1}{17}\right)$

c. Multiply $\dfrac{24}{11} \cdot \dfrac{9}{13}$

d. Multiply $\dfrac{x}{9} \cdot \dfrac{11}{y}$

e. Multiply $\dfrac{5x}{7} \cdot \dfrac{9y}{11}$

f. Multiply $\dfrac{x}{3} \cdot \dfrac{5y}{7} \cdot \dfrac{11z}{1}$

g. Multiply $\dfrac{3x}{y} \cdot \dfrac{z}{2a} \cdot \dfrac{5b}{c}$

h. Multiply $\dfrac{1}{x} \cdot \dfrac{3y}{5} \cdot \dfrac{7}{z} \cdot \dfrac{a}{b}$

i. Multiply $x \cdot y \cdot \dfrac{3}{z}$

MULTIPLYING FRACTIONS ANSWERS

a. $\dfrac{3 \cdot 7}{4 \cdot 13} = \dfrac{21}{52}$

b. $\dfrac{7 \cdot 1}{8 \cdot 17} = \dfrac{7}{136}$

c. $\dfrac{24 \cdot 9}{11 \cdot 13} = \dfrac{216}{143}$

FRACTIONS, ETC.

d. $\dfrac{x \bullet 11}{9 \bullet y} = \dfrac{11x}{9y}$

e. $\dfrac{5 \bullet x \bullet 9 \bullet y}{7 \bullet 11} = \dfrac{(5 \bullet 9) \bullet (x \bullet y)}{77} = \dfrac{45xy}{77}$

f. $\dfrac{x \bullet 5 \bullet y \bullet 11 \bullet z}{3 \bullet 7 \bullet 1} = \dfrac{(5 \bullet 11) \bullet (x \bullet y \bullet z)}{21} = \dfrac{55xyz}{21}$

g. $\dfrac{3 \bullet x \bullet z \bullet 5 \bullet b}{y \bullet 2 \bullet a \bullet c} = \dfrac{(3 \bullet 5) \bullet (x \bullet z \bullet b)}{2 \bullet (y \bullet a \bullet c)} = \dfrac{15bxz}{2acy}$

h. $\dfrac{1 \bullet 3 \bullet y \bullet 7 \bullet a}{x \bullet 5 \bullet z \bullet b} = \dfrac{(1 \bullet 3 \bullet 7) \bullet (y \bullet a)}{5 \bullet (x \bullet z \bullet b)} = \dfrac{21ay}{5bxz}$

i. $\dfrac{x}{1} \bullet \dfrac{y}{1} \bullet \dfrac{3}{z} = \dfrac{x \bullet y \bullet 3}{1 \bullet 1 \bullet z} = \dfrac{3xy}{z}$

Dividing Fractions

Dividing fractions is almost as easy as multiplying them.

To divide a fraction by another fraction, multiply by the reciprocal of the fraction you're dividing by (flip it upside-down or invert it).

If the division problem is written as a fraction, you'll invert the denominator and multiply it by the numerator.

CHAPTER 2

EXAMPLE

To divide one half by three fourths, $\dfrac{\left(\frac{1}{2}\right)}{\left(\frac{3}{4}\right)}$: Invert the bottom fraction and multiply it by the top fraction: $\left(\dfrac{1}{2}\right) \bullet \left(\dfrac{4}{3}\right)$. Then perform the multiplication: $\dfrac{1 \bullet 4}{2 \bullet 3} = \dfrac{4}{6}$, which simplifies to $\dfrac{2}{3}$.

To divide a whole number by a fraction, convert the whole number to a fraction (divide it by one) and then invert and multiply.

EXAMPLE

$\dfrac{x}{\left(\frac{3}{8}\right)}$ is the same as $\dfrac{\left(\frac{x}{1}\right)}{\left(\frac{3}{8}\right)}$. Now invert the bottom fraction and multiply: $\dfrac{x}{1} \bullet \dfrac{8}{3}$, or

To divide a fraction by a whole number, convert the whole number to a fraction (divide it by one) and then invert and multiply.

EXAMPLE

$\dfrac{\left(\frac{1}{2}\right)}{4}$ is the same as $\dfrac{\left(\frac{1}{2}\right)}{\left(\frac{4}{1}\right)}$. To divide, invert the denominator and multiply: $\left(\dfrac{1}{2}\right) \bullet \left(\dfrac{1}{4}\right)$. This is the same as $\dfrac{1 \bullet 1}{2 \bullet 4}$ or $\dfrac{1}{8}$.

FRACTIONS, ETC.

To divide more than one pair of fractions, work from the bottom of the expression up, inverting the denominator and multiplying each time.

EXAMPLE

To divide: $\dfrac{\left(\dfrac{1}{8}\right)}{\dfrac{\left(\dfrac{1}{2}\right)}{\left(\dfrac{1}{4}\right)}}$: First invert the bottom-most fraction $\left(\dfrac{1}{4}\right)$ and multiply it by the next fraction up, the one-half $\left(\dfrac{1}{2}\right)$: $\dfrac{\left(\dfrac{1}{8}\right)}{\left(\dfrac{1}{2}\right)\bullet\left(\dfrac{4}{1}\right)} = \dfrac{\left(\dfrac{1}{8}\right)}{\left(\dfrac{1\bullet 4}{2\bullet 1}\right)} = \dfrac{\left(\dfrac{1}{8}\right)}{\left(\dfrac{4}{2}\right)}$. Next, invert the denominator $\left(\dfrac{4}{2}\right)$ again and multiply by the numerator, one-eighth $\left(\dfrac{1}{8}\right)$:

$\left(\dfrac{1}{8}\right)\bullet\left(\dfrac{2}{4}\right) = \dfrac{1\bullet 2}{8\bullet 4} = \dfrac{2}{32}$, which simplifies to $\dfrac{1}{16}$.

All of these techniques work just as well when variables are substituted for numbers.

EXAMPLE

$\dfrac{\dfrac{y}{3}}{\dfrac{2}{x}}$ becomes $\dfrac{y}{3}\bullet\dfrac{x}{2}$ or $\dfrac{xy}{6}$

CHAPTER 2

DIVIDING FRACTIONS QUESTIONS

a. Divide $\dfrac{\left(\dfrac{3}{4}\right)}{\left(\dfrac{1}{2}\right)}$

b. Divide $\dfrac{7}{\left(\dfrac{3}{8}\right)}$

c. Divide $\dfrac{\left(\dfrac{4}{7}\right)}{5}$

d. Divide $\dfrac{\left(\dfrac{3}{7}\right)}{\left(\dfrac{\frac{3}{8}}{\frac{7}{9}}\right)}$

e. Divide $\dfrac{\left(\dfrac{a}{b}\right)}{c}$

f. Divide $\dfrac{\left(\dfrac{\frac{a}{b}}{\frac{c}{d}}\right)}{\left(\dfrac{\frac{e}{f}}{\frac{g}{h}}\right)}$

g. Divide $\dfrac{\left(\dfrac{x}{23}\right)}{17y}$

DIVIDING FRACTIONS ANSWERS

a. $\dfrac{3}{4} \bullet \dfrac{2}{1} = \dfrac{3 \bullet 2}{4 \bullet 1} = \dfrac{6}{4}$

b. $\dfrac{\left(\dfrac{7}{1}\right)}{\left(\dfrac{3}{8}\right)} = \left(\dfrac{7}{1}\right) \bullet \left(\dfrac{8}{3}\right) = \dfrac{7 \bullet 8}{1 \bullet 3} = \dfrac{56}{3}$

c. $\dfrac{\left(\dfrac{4}{7}\right)}{\left(\dfrac{5}{1}\right)} = \left(\dfrac{4}{7}\right) \bullet \left(\dfrac{1}{5}\right) = \dfrac{4 \bullet 1}{7 \bullet 5} = \dfrac{4}{35}$

d. $\dfrac{\left(\dfrac{3}{7}\right)}{\left(\dfrac{3}{8}\right) \bullet \left(\dfrac{9}{7}\right)} = \dfrac{\left(\dfrac{3}{7}\right)}{\dfrac{3 \bullet 9}{8 \bullet 7}} = \dfrac{\left(\dfrac{3}{7}\right)}{\left(\dfrac{27}{56}\right)} = \left(\dfrac{3}{7}\right) \bullet \left(\dfrac{56}{27}\right) = \dfrac{3 \bullet 56}{7 \bullet 27} = \dfrac{168}{189}$

e. $\dfrac{\left(\dfrac{a}{b}\right)}{\left(\dfrac{c}{1}\right)} = \left(\dfrac{a}{b}\right) \bullet \left(\dfrac{1}{c}\right) = \dfrac{a \bullet 1}{b \bullet c} = \dfrac{a}{bc}$

f. $\dfrac{\left(\dfrac{a}{b}\right) \bullet \left(\dfrac{d}{c}\right)}{\left(\dfrac{e}{f}\right) \bullet \left(\dfrac{h}{g}\right)} = \dfrac{\left(\dfrac{ad}{bc}\right)}{\left(\dfrac{eh}{fg}\right)} = \dfrac{ad}{bc} \bullet \dfrac{fg}{eh} = \dfrac{adfg}{bceh}$

g. $\dfrac{\left(\dfrac{x}{23}\right)}{\left(\dfrac{17y}{1}\right)} = \left(\dfrac{x}{23}\right) \bullet \left(\dfrac{1}{17y}\right) = \dfrac{x \bullet 1}{(23 \bullet 17) \bullet y} = \dfrac{x}{391y}$

Adding and Subtracting Fractions

SIGNS IN FRACTIONS: In a negative fraction, the minus sign may be placed in the numerator, the denominator, or before the fraction.

In other words, the position of the minus sign doesn't change the value of the fraction.

EXAMPLE

$\dfrac{-1}{3}$ is the same as $\dfrac{1}{-3}$, which is the same as $-\dfrac{1}{3}$.

Note, however, that one and only one minus sign must be used.

For example: $-\dfrac{1}{3} \neq \dfrac{-1}{-3}$.

To add numerical fractions, it is simplest to convert the fractions to decimal equivalents and then perform the addition.

EXAMPLE

$\dfrac{3}{8} + \dfrac{5}{7} = 0.375 + 0.7142857 = 1.0892857$ or approximately 1.09

Similarly, to subtract numerical fractions, convert to decimal equivalents and then subtract.

EXAMPLE

$$\frac{3}{4} - \frac{1}{2} = 0.75 - 0.5 = 0.25$$

If fractions contain variables, you must find a common denominator and then add the numerators.

If you can't use a calculator, knowing how to find a "least common denominator" is useful, but you will not need to know how to do this for certification exams; it is enough to be able to find a *common denominator* (not necessarily the *least* common denominator). Once fractions have the same denominator, you add the numerators to obtain the answer.

To find a common denominator, multiply each fraction by a fraction equal to one. Multiply the first fraction by a fraction whose *numerator and denominator* are the same as the *denominator* of the second fraction. (Since the numerator and denominator are the same, the fraction is equal to one, and multiplying by one does not change the value of a fraction). Then multiply the second fraction by a fraction (also equal to one) whose numerator and denominator are the same as the denominator of the first fraction. This is easier shown than said; review the following example:

EXAMPLE

To add $\frac{1}{x} + \frac{3}{8}$, first find a common denominator for the two fractions—in other words, make them have the same denominator without changing their values. Multiply the $\frac{1}{x}$ by $\frac{8}{8}$. The $\frac{8}{8}$ represents the denominator of the second fraction

17

$\left(\dfrac{3}{8}\right)$ divided by itself. When you perform this multiplication, we create: $\left(\dfrac{1}{x}\right) \bullet \left(\dfrac{8}{8}\right)$, which is the same as $\dfrac{1 \bullet 8}{x \bullet 8}$ or $\dfrac{8}{8x}$.

To create the second fraction's common denominator, multiply $\dfrac{3}{8}$ by $\dfrac{x}{x}$. The $\dfrac{x}{x}$ was created by dividing the denominator of the first fraction $\left(\dfrac{1}{x}\right)$ by itself. When you perform this multiplication, you get $\left(\dfrac{3}{8}\right) \bullet \left(\dfrac{x}{x}\right) = \dfrac{3 \bullet x}{8 \bullet x} = \dfrac{3x}{8x}$. Note that you have not changed the values of the fractions you're adding, but have simply written them in a different form.

Now that both fractions have the same denominator (8x), add the *numerators* only: $\left(\dfrac{8}{8x}\right) + \left(\dfrac{3x}{8x}\right) = \dfrac{8 + 3x}{8x}$. Remember that in addition and subtraction, unlike in multiplication and division, the 8 cannot be multiplied by the 3 in the 3x.

When *subtracting* fractions with variables, you must also find a common denominator. For example, $\dfrac{x}{3} - \dfrac{4}{7}$, a common denominator is 21 (seven times three): $\left(\dfrac{7}{7} \bullet \dfrac{x}{3}\right) - \left(\dfrac{3}{3} \bullet \dfrac{4}{7}\right)$ is the same as $\left(\dfrac{7 \bullet x}{7 \bullet 3}\right) - \left(\dfrac{3 \bullet 4}{3 \bullet 7}\right)$, which simplifies to $\dfrac{7x}{21} - \dfrac{12}{21}$ or $\dfrac{7x - 12}{21}$.

FRACTIONS, ETC.

To find a common denominator for more than two fractions, multiply every fraction by the "denominator-divided-by-the-denominator" of every other fraction.

EXAMPLE

The common denominator of $\frac{1}{2} + \frac{3}{x} - \frac{y}{5}$ must contain 2, x, and 5. The fractions multiplied to obtain a common denominator will be $\frac{2}{2}, \frac{x}{x}$, and $\frac{5}{5}$. To write the expression with a common denominator, multiply as follows:

$\left(\frac{x}{x} \cdot \frac{5}{5} \cdot \frac{1}{2}\right) + \left(\frac{2}{2} \cdot \frac{5}{5} \cdot \frac{3}{x}\right) - \left(\frac{2}{2} \cdot \frac{x}{x} \cdot \frac{y}{5}\right)$. This is the same as

$\left(\frac{x \bullet 5 \bullet 1}{x \bullet 5 \bullet 2}\right) + \left(\frac{2 \bullet 5 \bullet 3}{2 \bullet 5 \bullet x}\right) - \left(\frac{2 \bullet x \bullet y}{2 \bullet x \bullet 5}\right)$. Combine terms to get: $\frac{5x}{10x} + \frac{30}{10x} - \frac{2xy}{10x}$.

Now that there is a common denominator (10x) the numerators can be added and subtracted: $\frac{5x + 30 - 2xy}{10x}$.

ADDING AND SUBTRACTING FRACTIONS QUESTIONS

a. $\frac{1}{2} + \frac{3}{4} = ?$

b. $\frac{1}{2} - \frac{3}{4} = ?$

c. $\frac{3}{x} + \frac{1}{8} = ?$

d. $\frac{x}{7} + \frac{3}{8} = ?$

e. $\frac{7}{x} - \frac{4}{y} = ?$

f. $\frac{a}{b} + \frac{c}{d} = ?$

CHAPTER 2

g. $\dfrac{1}{8}+\dfrac{2}{5}-\dfrac{x}{y}=?$

h. $\dfrac{a}{b}-\dfrac{c}{d}-\dfrac{e}{f}=?$

ADDING AND SUBTRACTING FRACTIONS ANSWERS

a. $0.5+0.75 = 1.25$

b. $0.5-0.75 = -0.25$

c. $\left(\dfrac{8}{8}\bullet\dfrac{3}{x}\right)+\left(\dfrac{x}{x}\bullet\dfrac{1}{8}\right)=\left(\dfrac{8\bullet 3}{8\bullet x}\right)+\left(\dfrac{x\bullet 1}{x\bullet 8}\right)=\dfrac{24}{8x}+\dfrac{x}{8x}=\dfrac{24+x}{8x}$

d. $\left(\dfrac{8}{8}\bullet\dfrac{x}{7}\right)+\left(\dfrac{7}{7}\bullet\dfrac{3}{8}\right)=\left(\dfrac{8\bullet x}{8\bullet 7}\right)+\left(\dfrac{7\bullet 3}{7\bullet 8}\right)=\dfrac{8x}{56}+\dfrac{21}{56}=\dfrac{8x+21}{56}$

e. $\left(\dfrac{y}{y}\bullet\dfrac{7}{x}\right)-\left(\dfrac{x}{x}\bullet\dfrac{4}{y}\right)=\left(\dfrac{y\bullet 7}{y\bullet x}\right)-\left(\dfrac{x\bullet 4}{x\bullet y}\right)=\dfrac{7y}{xy}-\dfrac{4x}{xy}=\dfrac{7y-4x}{xy}$

f. $\left(\dfrac{d}{d}\bullet\dfrac{a}{b}\right)+\left(\dfrac{b}{b}\bullet\dfrac{c}{d}\right)=\left(\dfrac{d\bullet a}{d\bullet b}\right)+\left(\dfrac{b\bullet c}{b\bullet d}\right)=\dfrac{ad}{bd}+\dfrac{bc}{bd}=\dfrac{ad+bc}{bd}$

g. $\left(\dfrac{5}{5}\bullet\dfrac{y}{y}\bullet\dfrac{1}{8}\right)+\left(\dfrac{8}{8}\bullet\dfrac{y}{y}\bullet\dfrac{2}{5}\right)-\left(\dfrac{8}{8}\bullet\dfrac{5}{5}\bullet\dfrac{x}{y}\right)=\dfrac{5y}{40y}+\dfrac{16y}{40y}-\dfrac{40x}{40y}=\dfrac{5y+16y-40x}{40y}=\dfrac{21y-40x}{40y}$

h. $\left(\dfrac{d}{d}\bullet\dfrac{f}{f}\bullet\dfrac{a}{b}\right)-\left(\dfrac{b}{b}\bullet\dfrac{f}{f}\bullet\dfrac{c}{d}\right)-\left(\dfrac{b}{b}\bullet\dfrac{d}{d}\bullet\dfrac{e}{f}\right)=\dfrac{adf}{bdf}-\dfrac{bcf}{bdf}-\dfrac{bde}{bdf}=\dfrac{adf-bcf-bde}{bdf}$

Reciprocals

Reciprocals are two numbers whose product is one.

EXAMPLE

$\frac{3}{4}$ and $\frac{4}{3}$ are reciprocals because when they are multiplied, the product $\left(\frac{12}{12}\right)$ equals one.

The reciprocal of any whole number or expression is 1 divided by that whole number or expression.

EXAMPLE

$3 \cdot \frac{1}{3}$ is the same as $\frac{3}{1} \cdot \frac{1}{3} = \frac{3 \cdot 1}{1 \cdot 3}$. This equals $\frac{3}{3}$ or 1. This means that the reciprocal of three is one third. Using variables, $x \cdot \frac{1}{x}$ is the same as $\frac{x}{1} \cdot \frac{1}{x}$. Performing the multiplication: $\frac{x}{x} = 1$. So the reciprocal of x is $\frac{1}{x}$. The same is true for complex expressions. For example, $(x+3) \cdot \frac{1}{(x+3)} = \frac{x+3}{x+3} = 1$. Therefore, the reciprocal of x + 3 is $\frac{1}{x+3}$.

The reciprocal of a fraction is the inverse of the fraction.

(The numerator becomes the denominator and vice versa).

EXAMPLE

The reciprocal of $\frac{3}{8}$ is $\frac{8}{3}$ because the product $\left(\frac{24}{24}\right)$ is equal to one.

This is also true for complex fractions.

EXAMPLE

The reciprocal of $\dfrac{3x+6}{4y-5}$ is the inverse: $\dfrac{4y-5}{3x+6}$.

The term "reciprocal" is often used in word problems, and must be understood in order to complete the safety certification examinations. For example, the test phrase "the reciprocal of a distance, 5-miles" would be expressed as $\dfrac{1}{5(miles)}$. Many scientific calculators have a reciprocal key, usually marked "$1/x$" or "x^{-1}."

Proportions

A proportion exists when the cross-products of two fractions are equal.

Cross-products of two fractions are found by cross-multiplying the numerator of the first fraction by the denominator of the second, and the denominator of the first fraction by the numerator of the second.

EXAMPLE

$\dfrac{3}{4} = \dfrac{90}{120}$ is a proportion because $3 \bullet 120 = 360$ and $4 \bullet 90 = 360$ Since the products (360) are equal, the equation is called a proportion.

RECIPROCAL AND PROPORTION QUESTIONS

a. Are $\dfrac{17}{256}$ and $\dfrac{256}{17}$ reciprocals?

b. Are $\dfrac{17}{51}$ and $\dfrac{21}{7}$ reciprocals?

FRACTIONS, ETC.

c. Are $\dfrac{1}{8}$ and 8 reciprocals?

d. Are $\dfrac{4}{28}$ and 7 reciprocals?

e. What is the reciprocal of $\dfrac{4}{x}$?

f. What is the reciprocal of $2x+9$?

g. Is $\dfrac{7}{16} = \dfrac{35}{80}$ a proportion?

h. Is $\dfrac{9}{78} = \dfrac{4}{65}$ a proportion?

i. 28 moles of a gaseous mixture can be purchased for $7, or 42 moles for $10.50. Are all moles the same price?

RECIPROCAL AND PROPORTION ANSWERS

a. Yes, because their product is one: $\dfrac{17}{256} \bullet \dfrac{256}{17} = \dfrac{4{,}352}{4{,}352} = 1$

b. Yes, because their product is one: $\dfrac{17}{51} \bullet \dfrac{21}{7} = \dfrac{357}{357} = 1$

c. Yes, because their product is one: $\dfrac{1}{8} \bullet \dfrac{8}{1} = \dfrac{8}{8} = 1$

d. Yes, because their product is one: $\dfrac{4}{28} \bullet \dfrac{7}{1} = \dfrac{28}{28} = 1$

e. The reciprocal is $\dfrac{x}{4}$, because when $\dfrac{x}{4}$ multiplies $\dfrac{4}{x}$, the product is one:

$\dfrac{4}{x} \bullet \dfrac{x}{4} = \dfrac{4x}{4x} = 1$

f. The reciprocal is $\dfrac{1}{2x+9}$ because when these are multiplied, the product is one:
$$\dfrac{2x+9}{1} \cdot \dfrac{1}{2x+9} = \dfrac{2x+9}{2x+9} = 1$$

g. Yes, because the products of cross-multiplication are equal: $7 \bullet 80 = 560$ and $16 \bullet 35 = 560$

h. No, because the products of cross-multiplication are not equal: $9 \bullet 65 = 585$ but $78 \bullet 4 = 312$ and $585 \neq 312$

i. Yes, because the products of cross multiplication are equal:
$$\dfrac{Moles \rightarrow}{Dollars \rightarrow} \quad \dfrac{28}{7} = \dfrac{42}{10.5} \quad \dfrac{\leftarrow Moles}{\leftarrow Dollars} \qquad 28 \bullet 10.5 = 294 \text{ and } 7 \bullet 42 = 294$$

Rounding

Significant digits: The "most significant digit" is the first non-zero digit from the left for any number written in decimal form.

This is true even if the number is a whole number and has no decimal places. All digits to the right of the most significant digit are also "significant." The number of significant digits is found by counting from the "most significant digit" to the right.

EXAMPLE

In 128.00, one (the leftmost non-zero digit) is the most significant digit, and there are five significant digits (the one plus all to its right). Since the zeros are written, and to the right of the most significant digit, they also become significant. Had the number been written as 128, there would only be three significant digits.

For another example, in 0.00208, two (the leftmost non-zero digit) is the most significant digit, and there are three significant digits (the two plus all of the written digits to its right). The three zeros that precede the two are not significant because they are to the left of the "most significant digit."

In any type of math problem, the number with the smallest number of digits to the right of the decimal point determines the number of decimal places used in the answer.

When multiplying a whole number (with no decimal places shown) and a fraction, the answer is never more accurate than a whole number. In order to maintain accuracy, don't round until you have a final answer. This eliminates "calculator error" in determining the best answer.

EXAMPLE

284 • 0.55 gives a calculator answer of 156.2, but rounds to 156 because the least accurate number in the problem (the 284) determines the number of decimal places used in the answer. If, on the other hand, the problem is written as 284.0 • 0.55, the correct answer is 156.2. If the problem is written as 284.00 • 0.55, the correct answer is 156.20.

Note that problems that contain unit conversions cannot be used to determine the number of decimal places in the answer. For example, if one foot converts to twelve inches (exactly), the inch unit is the same as 12.000... to an infinite number of decimal places.

When rounding to a specific number of significant digits, use the next digit to the right to determine whether to round up or down.

EXAMPLE

To round the number 28.563 to four significant digits (two decimal places), the next number to the right (three) is the used for rounding. To round the same number to three significant digits (one decimal place), the next number to the right (six) is used for rounding.

When determining whether to round up or down, use these rules:

25

CHAPTER 2

If the digit used for rounding is four or less, discard it and let the prior digit stand as is. This is frequently called rounding down.

EXAMPLE

To round 28.563 to four significant digits (two decimal places), the digit used for rounding is the third decimal place: three. Because the three meets the "four or less" criteria, the prior digit stands, and the answer is 28.56.

If the digit used for rounding is six or more, round the prior digit up.

EXAMPLE

To round 28.563 to three significant digits (one decimal place), the digit used for rounding is the second decimal place: six. Because six meets the "six or more" criteria, the previous digit (five) is rounded up to a six, and the answer is 28.6.

If the digit used for rounding is exactly five, determine how to round by looking at the digit to the left of the digit used for rounding: If it is even, leave it as it is; if it is odd, round it up to the next even digit.

EXAMPLE

To round 28.5 to two significant digits (no decimal places), the digit used for rounding is five. Since five meets the "exactly five" criteria, look at the next digit to its left. Since the eight is already an even number, the number rounds down to 28. If the number had been 27.5, it would have rounded up to 28.

The number of significant digits asked for in the answer can determine the number used for rounding.

EXAMPLE

To round 28.563 to two significant digits (no decimal places), the fact that you are asked for two significant digits means that *only the five is used for rounding*. This means that the number to the left (the eight) and NOT the numbers to the right are significant in determining the answer. Because the number to the left (the eight) is

FRACTIONS, ETC.

even, this number rounds to 28 (and not 29), even though the ".563" is more than halfway to the next digit.

ROUNDING QUESTIONS

a. $28.5 \bullet 1.436 = ?$

b. Round 33.2895 to two significant digits.

c. Round 33.2895 to three significant digits.

d. Round 33.2895 to four significant digits.

e. Round 33.2895 to five significant digits.

ROUNDING ANSWERS

a. 40.9 because the accuracy of the answer is limited to one decimal place by the first number being multiplied.

b. 33 because the digit used for rounding (2) is "four or less."

c. 33.3 because the digit used for rounding (8) is "six or more" so the next digit to the left (2) rounds up.

d. 33.29 because the digit used for rounding (9) is "six or more" so the next digit to the left (8) rounds up.

e. 33.290 because the digit used for rounding (5) is "exactly five," the next digit to the left (9) is odd, and therefore rounds up to 10. Since five significant digits are required, the final digit becomes a zero.

Absolute Value

Absolute value is used in some mathematical calculations on safety certification examinations, particularly in the NIOSH lifting equation. Absolute value symbols are upright lines with expressions inside: $|7y-x|$, $\left|\frac{3}{y}-5\right|$, and $|V-30|$ are all examples of absolute values. To determine an absolute value, first work the expression between the absolute value symbols. If the value of the expression is positive, that answer is

27

CHAPTER 2

the absolute value. If the value of the expression is negative, delete the minus sign, and what is left is the absolute value.

EXAMPLE

The absolute value of $|7-4|$ is 3. The absolute value of $|7-10|$ is not -3, but also a positive 3.

ABSOLUTE VALUE QUESTIONS

a. Is this a true equation? $28 - 24 = |24 - 28|$

b. Is this a true statement? $|-x+5| = x+5$

c. What is the absolute value of this expression? $|28 + (4-14)|$

ABSOLUTE VALUE ANSWERS

a. Yes, because $28 - 24 = 4$ and $|24 - 28| = 4$

b. No, because the value of $(-x+5)$ can be positive or negative, depending on the value of x. For example, if $x = 20$, then $(-x+5) = -15$, and the absolute value of the expression is 15. This is not the same as $x + 5$, which would be 25, in this case.

c. 18, because the expression simplifies to 28-10. Since the result (18) is already positive, it is the absolute value.

FRACTIONS, ETC. CHALLENGE EXAM

a. Convert $\dfrac{726}{816}$ to both decimal and percentage forms.

b. Convert 0.0028% to a decimal form.

c. $\dfrac{28x}{5} \bullet \dfrac{33}{17yz}$ simplifies to what?

d. $167 \bullet \dfrac{43x}{9y}$ simplifies to what?

FRACTIONS, ETC.

e. $\dfrac{2a}{b} \bullet 3 \bullet \dfrac{7c}{5d}$ simplifies to what?

f. $\dfrac{\left(\dfrac{3y}{4x}\right)}{\dfrac{\left(\dfrac{z}{5}\right)}{\left(\dfrac{7}{9}\right)}}$ simplifies to what?

g. $\dfrac{7x}{8y} + \dfrac{1}{3z}$ simplifies to what?

h. $\dfrac{2x}{y} + \dfrac{3z}{5} - \dfrac{b}{7a}$ simplifies to what?

i. What is the reciprocal of $\dfrac{27y}{36x}$?

j. Find the reciprocal of 28zx-13yz+17x.

k. Is the expression $\dfrac{28}{17x} = \dfrac{256}{x}$ a proportion?

l. Round 28.56859 to five significant digits.

m. Round the answer: $28.3 \bullet 1.2809151$.

n. What is the absolute value of this expression: $|28 - (7 \bullet 5)|$?

FRACTIONS, ETC. CHALLENGE EXAM ANSWERS

a. Approximately 0.89 (decimal) and 89% (percent)

b. 0.000028

c. $\dfrac{924x}{85yz}$

d. $\dfrac{7181x}{9y}$

29

e. $\dfrac{42ac}{5bd}$

f. $\dfrac{105y}{36xz}$

g. $\dfrac{21xz + 8y}{24yz}$

h. $\dfrac{70ax + 21ayz - 5by}{35ay}$

i. $\dfrac{36x}{27y}$

j. $\dfrac{1}{28zx - 13yz + 17x}$

k. No, because $28x \neq 4{,}352x$

l. 28.568, because the digit used for rounding is exactly five, the next digit to the *left* (the eight) is already even, and does not change.

m. 36.2, because the answer can't be accurate to more than one decimal place.

n. 7, because the expression becomes: $28 - 35 = -7$. $|-7| = 7$

CHAPTER THREE

Exponents, Roots, and Logs

> **Assumptions:** You know how to add, subtract, multiply, and divide. You have a scientific calculator (and basic knowledge of how to use it). You can multiply fractions, and know how to round numbers.
>
> **Application:** Required for ASP, CSP, OHST, and CHST examinations.
>
> **Discussion:** Exponents, roots, and logarithms are "scientific shorthand." These concepts are used extensively in all certification examinations. Some descriptions are included with calculator manuals, but this chapter is intended to provide background and elaboration. If you feel your comprehension of exponents, roots, and logarithms is good, go to the end of this chapter and try the challenge exam. If you can work all problems correctly, skip this chapter.

What Is an Exponent?

Exponents are a "shorthand" way to describe repeated instances of multiplying numbers by themselves. When you multiply numbers by themselves, you sometimes write out the sequence. For example: $3 \bullet 3$. This is feasible if only one or two multiplications are to be performed, but it becomes cumbersome if you must write: $3 \bullet 3 \bullet 3 \bullet 3 \bullet 3 \bullet 3$.

Exponents are a way to represent how many times a number multiplies itself.

$3 \bullet 3$ is represented with exponents by putting a smaller number to the upper right of the number being multiplied by itself. Thus, $3 \bullet 3$ becomes 3^2. In this example, the number multiplying itself (the three) is called the "base," and the number of times the base occurs in the multiplication sequence (two) is called the "exponent." The exponent merely describes how many times the base is multiplied by itself. Since the exponent in this example is a two, one three is multiplied by another three.

To express $3 \bullet 3 \bullet 3 \bullet 3 \bullet 3 \bullet 3$ in exponential form, count the threes. Since there are six threes, the exponent is six, and the base is three. This is written 3^6. In both cases ($3 \bullet 3 \bullet 3 \bullet 3 \bullet 3 \bullet 3$ and 3^6) the value is the same (three times three = 9, times three = 27, times three = 81, times three = 243, times three = 729).

Rules of Exponents

Any base to the power of one is the base itself.

EXAMPLE

$10^1 = 10$, $3^1 = 3$, and $x^1 = x$. This works regardless of what the base is. If the exponent is one, the value of the expression is always the same as the base.

Any non-zero base to the power of zero is equal to one.

EXAMPLE

Both 10^0 and 3^0 are equal to 1. This works for all non-zero bases. If the exponent is zero, the value of the expression is one.

A negative exponent is the same as the reciprocal of the base to the positive value of the exponent.

EXAMPLE

3^{-1} is the same as $\frac{1}{3^1}$ or $\frac{1}{3}$. 3^{-2} is the same as $\frac{1}{3^2}$ or $\frac{1}{3 \bullet 3}$ or $\frac{1}{9}$. Similarly, x^{-3} is the same as $\frac{1}{x^3}$. Note that the exponent becomes positive when you form the reciprocal.

This also works if the negative exponent is in the denominator of the fraction. The reciprocal of the expression with the negative exponent in the denominator has a positive exponent.

EXPONENTS, ROOTS AND LOGS

EXAMPLE

$\frac{1}{3^{-1}}$ is the same as $\frac{3^1}{1}$ which becomes 3^1 which simplifies to just 3.

Also, $\frac{1}{x^{-2}}$ is the same as $\frac{x^2}{1}$ which simplifies to x^2.

To multiply two expressions with the same base, add the exponents.

EXAMPLE

$3^2 \cdot 3^2$ is equal to 3^{2+2} or 3^4. You can verify this by working the sequence both ways. $3^2 \cdot 3^2 = 9 \cdot 9 = 81$, while $3^4 = 3 \cdot 3 \cdot 3 \cdot 3 = 9 \cdot 3 \cdot 3 = 27 \cdot 3 = 81$. This is true for any non-zero base. Using variables: $x^i \cdot x^j \cdot x^k = x^{i+j+k}$

Note that in all these sequences, *the bases must be the same*, and *the expressions must be multiplied by each other* (not added or subtracted from each other).

EXAMPLE

$3^2 \cdot 2^3 \neq 5^5$ nor is it equal to 6^5 (the bases must be the same). Also, $3^2 + 3^3 \neq 3^5$.

The expressions must be multiplied together, not added or subtracted to use this rule.

To divide two expressions with the same base, subtract their exponents.

EXAMPLE

$\frac{3^5}{3^2}$ is equal to 3^{5-2} or 3^3.

Note that the *bases must be the same, the expressions must be divided by each other,* and *the exponent in the denominator is subtracted from the exponent in the numerator*. This is true for all non-zero bases: $\frac{x^a}{x^b} = x^{a-b}$.

CHAPTER 3

When a product is raised to a power, all of the exponents in the product are multiplied by that power.

For example, $(2 \bullet 4)^2$ is the same as $2^2 \bullet 4^2$. This can be verified by working the expression both ways: $(2 \bullet 4)^2 = 8^2 = 64$. This is the same as $2^2 \bullet 4^2 = 4 \bullet 16 = 64$. This does *not* hold true if the bases in the expression are added or subtracted.

EXAMPLE

$(3^2 - 4^3)^4 \neq 3^{2 \bullet 4} - 4^{3 \bullet 4}$ because:
$3^{2 \bullet 4} - 4^{3 \bullet 4} = 3^8 - 4^{12} = 6,561 - 16,777,216 = -16,770,655$,
while $(3^2 - 4^3)^4 = (3^2 - 4^3)^4 = (9 - 64)^4 = -55^4 = 9,150,625$

(the correct answer). This is true for all bases multiplied together: $\left(\dfrac{x \bullet 3y}{c}\right)^z$ is equal to $\dfrac{x^z \bullet 3^z \bullet y^z}{c^z}$ Note that every unit (variable or number) in the expression is raised to the z power whether it is in the numerator or denominator. This means that $\left(\dfrac{x}{y}\right)^z = \dfrac{x^z}{y^z}$.

By common convention, any base raised to the power of two is said to be "squared." Any base raised to the power of three is said to be "cubed." All other powers are given their numerical name. For example the expression x^7 is pronounced "x to the seventh power."

EXPONENT QUESTIONS

a. Write $3 \bullet 3 \bullet 3$ in exponent form.

b. Write x^4 without exponents.

c. $278^1 = ?$

d. $278^0 = ?$

e. Write 28^{-4} using only positive exponents.

EXPONENTS, ROOTS AND LOGS

f. Write $\dfrac{28}{x^{-4}}$ using only positive exponents.

g. Simplify: $x^2 \bullet x^2 \bullet x^3$

h. Simplify: $\dfrac{x^{17}}{x^{13}}$

i. Simplify: $(3x \bullet 4y^3)^2$

EXPONENT ANSWERS

a. 3^3

b. $x \bullet x \bullet x \bullet x$

c. 278

d. 1

e. $\dfrac{1}{28^4}$

f. $28x^4$ derived from: $\dfrac{\left(\dfrac{28}{1}\right)}{\left(\dfrac{x^{-4}}{1}\right)} = \left(\dfrac{28}{1}\right) \bullet \left(\dfrac{x^4}{1}\right) = \dfrac{28 \bullet x^4}{1 \bullet 1} = 28x^4$

g. x^7, because $x^2 \bullet x^2 \bullet x^3 = x^{2+2+3} = x^7$

h. x^4, because $\dfrac{x^{17}}{x^{13}} = x^{17-13} = x^4$

i. $144x^2y^6$, because $(3x \bullet 4y^3)^2 = (3^2 \bullet x^2) \bullet (4^2 \bullet y^{3 \bullet 2}) = 9 \bullet x^2 \bullet 16 \bullet y^6 = 144x^2y^6$

What Is a Root?

A root is the opposite of an exponent.

Exponents indicate how many times a base is multiplied by itself. A root provides the product and asks what base must be multiplied by itself a specific number of times to obtain that product.

In talking about roots, the following terminology is used. The "root symbol" ($\sqrt{}$) is called the "radical." The number inside the radical $\sqrt[3]{125}$ (125 in this case) is called the "radicand." The number on the outside of the radical $\sqrt[3]{125}$ (three, in this case) is called the "order." If no order is shown, the order is assumed to be two (a "square root"). You don't need to memorize these terms for the Safety Certification exams, but they will be used in this chapter to simplify the discussion of roots.

The expression shown here "the square root of four," ($\sqrt{4}$) asks "What number, when squared, equals 4?" The symbol $\sqrt{}$ means "the square root of." Obviously, two squared ($2 \bullet 2$ or 2^2) is four; so a square root of four is two.

Even-order roots can be positive or negative.

Interestingly, in the example above, two is not the only square root of four. Negative two is also a square root of four, because when a negative two is multiplied by negative two, positive four is the product. This means that an even-order root (2,4,6, etc.) of a radicand can be either positive or negative in value. It is thus true to say that the square roots of four include both positive and negative two (± 2). There will always be two roots (a positive and a negative) for an even-order radical.

Even-order roots of negative radicands are imaginary numbers.

There is no real-number root (positive or negative) that can provide a negative number when multiplied by itself an even number of times. This means that even-order roots of negative numbers can only be imaginary numbers (not used in safety certification exams). There is no real-number square root of negative 25, for example ($\sqrt{-25}$) because neither positive five nor negative five produces a negative 25 when squared.

EXPONENTS, ROOTS AND LOGS

Orders can be higher than two.

EXAMPLE

The third (cube) root of 27 ($\sqrt[3]{27}$) asks "What number, multiplied by itself three times, equals 27?"

When orders are odd (3,5,7, etc.), there can be only one root, and the root will retain the sign of the radicand. If the cube root of 27 is 3 (three times three is nine, times three is twenty-seven), then the cube root of negative twenty-seven (-27) is negative three (-3). This is true because negative three times negative three gives positive nine, times negative three is negative 27.

To take the sixth root of 729, ($\sqrt[6]{729}$), have your calculator do the math. Most calculators have a "universal root key" ($\sqrt[y]{x}$ or $\sqrt[x]{y}$). The key sequence varies among brands of calculators. Most calculators give only the numerical value of the root (3) and not the signs of both roots (±3) if there are two roots.

Roots can be expressed as exponents.

The expression "the square root of nine" ($\sqrt{9}$) is the same as taking the radicand (nine) to a fractional power.

EXAMPLE

$\sqrt{9}$ is the same as $9^{\left(\frac{1}{2}\right)}$. The radicand (nine) in the square root radical became the base of the expression with the fractional exponent. The order of the radical (two, since this was a square root), becomes the denominator of the exponent, and the exponent of the radicand (one, in this case, since the nine inside the radical is not raised to any power beyond one), becomes the numerator of the exponent. For another example, $\sqrt[7]{-256}$ is the same as $-256^{\left(\frac{1}{7}\right)}$.

CHAPTER 3

All of the rules of exponents apply to fractional exponents.

Note that the when an expression with a fractional exponent is raised to a power, the exponents are multiplied together, just as if they had been whole numbers.

EXAMPLE

$\left(x^{\frac{1}{3}}\right)^2$ is the same as $x^{\left(\frac{1}{3} \cdot \frac{2}{1}\right)} = x^{\left(\frac{2}{3}\right)}$. This could also be expressed as $\sqrt[3]{x^2}$. Note that the denominator of the exponent becomes the order of the radical, and the numerator of the exponent becomes the exponent of the radicand inside. Another example, $\sqrt[9]{y^7}$ is the same as $y^{\left(\frac{7}{9}\right)}$.

ROOTS QUESTIONS

a. Calculate: $\sqrt{64}$

b. Calculate: $\sqrt[3]{512}$

c. Calculate: $\sqrt[3]{-125}$

d. Write $\sqrt[8]{1{,}750}$ using exponents.

e. Write $x^{\left(\frac{4}{5}\right)}$ using a radical.

f. Multiply exponents and write $\left(x^2 \cdot x^5\right)^{\frac{1}{8}}$ using a radical.

ROOTS ANSWERS

a. ± 8, because $8 \cdot 8 = 64$ and $(-8) \cdot (-8) = 64$

b. 8, because $8 \cdot 8 \cdot 8 = 512$ (-8 is NOT a root because $(-8) \cdot (-8) \cdot (-8) = -512$

c. -5, because $(-5) \cdot (-5) \cdot (-5) = -125$

EXPONENTS, ROOTS AND LOGS

d. $1{,}750^{\left(\frac{1}{8}\right)}$

e. $\sqrt[5]{x^4}$

f. $\sqrt[8]{x^7}$, because $(x^2 \bullet x^5)^{\frac{1}{8}} = (x^{2+5})^{\frac{1}{8}} = (x^7)^{\frac{1}{8}} = x^{\left(\frac{7}{1}\right)\bullet\left(\frac{1}{8}\right)} = x^{\frac{7\bullet 1}{1\bullet 8}} = x^{\left(\frac{7}{8}\right)} = \sqrt[8]{x^7}$

Logarithms and Antilogs

A "common" logarithm is an exponent of the base ten.

The base of all common logarithms or "logs"(unless a logarithm is called a "natural log," it is assumed to be a common log) is ten. Since $10^1 = 10$ and $10^2 = 100$, the log of 10 is one (the same as the exponent), and the log of 100 is two (the same as the exponent).

This works fine if the number you're seeking a log of is an even multiple of ten. What happens if you want the log of 56? The exponent of ten that will produce 56 must obviously be more than one (which would give you 10), but less than two (which would give you 100). Some fraction's decimal equivalent between one and two must be the log of 56. Fortunately, scientific calculators have a button that provides logarithms without forcing you to do the calculations. Entering 56 and then pushing that button (labeled LOG on most calculators) provides the answer: 1.748188027. This means that $10^{1.748188027} = 56$. You can also say "the log of 56 is 1.748188027." The log of a number between one (10^0) and ten (10^1) will be between 0 and 1. For example, the log of 6 is 0.778, approximately.

Logs can be negative numbers.

As you saw above, $10^1 = 10$. It is also true that $10^0 = 1$. If the log of one is zero, what would be the log of a number less than one? As it turns out, ten to a negative exponent provides values less than one.

EXAMPLE

10^{-4} would be 0.0001. Thus, the log of one ten-thousandth (0.0001) is negative four. Another example: The log of 0.004827 is -2.316, approximately. This is the same is saying that $10^{-2.31622701} = 0.004827$.

CHAPTER 3

An antilog is the opposite of a log.

A logarithm is an exponent; an antilog is what you get if you raise ten to a power of that exponent.

EXAMPLE

The antilog of two is 100, because $10^2 = 100$. In effect, an antilog **gives you** the exponent (the log), and asks what the product is if you raise ten to that power. The antilog of -1.28746 would be 0.051586968 because $10^{-1.28746} = 0.51586968$. Scientific calculators usually perform the common antilog function with a key labeled 10^x.

"Natural" logs use base e.

The only other type of log used on the safety certification examinations is called a "natural" log. Natural logs use the base "e" instead of the base ten. "e" stands for the number 2.718281828.... This number is derived so that on an x-y graph, a change in the y value produces an equivalent increase in the area under a curve. You need not know the derivation of "e" nor the value of it. The "e" button on the scientific calculator (sometimes marked LN) provides this number. Just as a common log is the value of the exponent on base ten needed to produce a specific value, the natural log represents the value of an exponent on a base of 2.718281828 (e) needed to produce that value.

EXAMPLE

The natural log of 10 is 2.302585093. This means that e to the power of 2.302585093 ($2.718281828^{2.302585093}$) is equal to ten. As with common logs, negative natural logs represent values less than one. So $e^{-2.48765}$ would equal 0.083105034. This is the same as saying "the natural log of 0.83105034 is -2.48765."

An anti-natural log is the opposite of a natural log.

A natural log is an exponent; the anti-natural log is the product you get if you raise e to that exponent.

EXPONENTS, ROOTS AND LOGS

EXAMPLE

The anti-natural log of two is 7.389056099 because $e^2 = 2.718281828^2 = 7.389056099$. The anti-natural log *gives you* the exponent (the natural log), and asks what the result would be if you raised e to that particular value. The anti-natural log key on most scientific calculators is labeled "e^x".

LOG AND ANTILOG QUESTIONS

a. The log of 1,000 = ?
b. The common log of 27 = ?
c. What is the log of 0.000276?
d. The antilog of 2.5 = ?
e. The common antilog of -2.5 = ?
f. The natural log of 2.718281828 = ?
g. The natural log of 0.093 = ?
h. The anti-natural log of 5.3 = ?

LOG AND ANTILOG ANSWERS

a. 3, because $10^3 = 10 \cdot 10 \cdot 10 = 1,000$
b. 1.431363764, because $10^{1.431363764} = 27$
c. -3.559090918, because $10^{-3.559090918} = 0.000276$
d. 316.227766, because $10^{2.5} = 316.227766$
e. 0.003162278, because $10^{-2.5} = 0.003162278$
f. 1, because $e^1 = e = 2.718281828$
g. Approximately -2.375, because $e^{-2.375155786} = 2.718281828^{-2.375155786} = 0.093$
h. 200.33681, because $e^{5.3} = 2.718281828^{5.3} = 200.33681$

CHAPTER 3

EXPONENTS, ROOTS, AND LOGS CHALLENGE EXAM

a. Write $7 \cdot 7 \cdot 7 \cdot 7 \cdot 7 \cdot 7$ using exponents.

b. What is the value of $(7x)^1$?

c. What is the value of $(2,728,046)^0$?

d. Write x^{-7} using only positive exponents.

e. Simplify: $x^5 \cdot x^{-3}$

f. Simplify: $\dfrac{x^5}{x^{-3}}$

g. Simplify: $\left(\dfrac{x^2}{y^3}\right)^{2z}$

h. $\sqrt{256} = ?$

i. $\sqrt[3]{-64x^3} = ?$

j. $\sqrt{-100} = ?$

k. Express $\sqrt[z]{x^y}$ without using a radical.

l. Multiply exponents and write $\left(x^3 \cdot x^{-5}\right)^{\frac{3}{8}}$ using a radical and only positive exponents.

m. Log 365 = ?

n. Common antilog 4.28 = ?

o. Natural log of 698 = ?

p. Anti-natural log of 4 = ?

EXPONENTS, ROOTS, AND LOGS CHALLENGE EXAM ANSWERS

a. 7^6, because there are six sevens in the original multiplication.

b. 7x, because $(7x)^1 = 7^1 \cdot x^1 = 7x$

EXPONENTS, ROOTS AND LOGS

c. 1, because *any* non-zero root to the zero power is equal to one.

d. $\dfrac{1}{x^7}$, because the reciprocal of the base $\dfrac{1}{x}$ allows the exponent to change value $\dfrac{1}{x^7}$.

e. x^2, because when bases multiply, exponents add. 5+(-3) = 2.

f. x^8, because when bases are divided, exponents subtract. 5-(-3) = 5+3 = 8.

g. $\dfrac{x^{4z}}{y^{6z}}$, because all exponents inside the parentheses are multiplied by the exponent outside the parentheses: $\dfrac{x^{2 \cdot 2z}}{y^{3 \cdot 2z}} = \dfrac{x^{4z}}{y^{6z}}$

h. ± 16, because $16 \bullet 16 = 256$ and $-16 \bullet -16 = +256$

i. $-4x$, because $\sqrt[3]{-64x^3} = \left(\sqrt[3]{-64}\right) \bullet \left(\sqrt[3]{x^3}\right) = -4 \bullet x = -4x$

j. No real number, because even orders can never produce a negative radicand.

k. $x^{\frac{y}{z}}$, because the order (z) becomes the denominator of the exponent, and the power of the radicand (y) becomes the numerator.

l. $\sqrt[4]{\dfrac{1}{x^3}}$, because $\left(x^3 \bullet x^{-5}\right)^{\frac{3}{8}} = \left(x^{3+(-5)}\right)^{\frac{3}{8}} = \left(x^{-2}\right)^{\frac{3}{8}} = x^{\frac{-2}{1} \bullet \frac{3}{8}} = x^{\frac{-6}{8}} = x^{\frac{-3}{4}} = \sqrt[4]{x^{-3}} = \sqrt[4]{\dfrac{1}{x^3}}$

m. 2.562292864, because $10^{2.562292864} = 365$

n. 19,054.60718, because $10^{4.28} = 19,054.60718$

o. 6.548219103, because $e^{6.548219103} = 2.718281828^{6.548219103} = 698$

p. 54.59815003, because $e^4 = 2.718281828^4 = 54.59815003$

CHAPTER FOUR

Systems of Measurement

> **Assumptions:** You know how to add, subtract, multiply, and divide. You have a scientific calculator (and basic knowledge of how to use it). You can multiply fractions, and you know how to round numbers.
>
> **Application:** Required for ASP, CSP, OHST, and CHST examinations.
>
> **Discussion:** English and metric systems of measurement are presumed by the BCSP to be "relatively simple" material that candidates are expected to know. You must memorize units and their scaling factors to pass the Safety Certification examinations. If you feel that your comprehension of measurement systems is good, go to the end of this chapter and try the challenge exam. If you can work all of the problems correctly, skip this chapter.

English Units

English units of measuring weight, volume, liquids, time, distance, and temperature are probably familiar to you. Each type of measurement has a base unit (miles, gallons, etc.) with sub-units (yards, pints, etc.). These units will appear frequently in safety certification examinations. You will be expected to know the relationships between these units of measure.

Weight – The Pound

The primary unit of weight measure in the English system is the pound. A pound is subdivided by 16 ounces. These are ounces of *weight* as opposed to ounces of liquid measure.

Volume – The Cubic Foot

The primary unit of volume measure in the English system is the cubic foot. A cubic foot is divided into 1,728 cubic inches. This number comes from multiplying the 12 inches of height by the 12 inches of width by the 12 inches of depth in a one-foot cube.

Liquids – The Gallon

The primary unit of liquid volume in the English system is the gallon. A gallon can be subdivided into half-gallons, quarts, pints, cups, and ounces. It can also be divided into fifths, as is common when selling alcoholic beverages. One gallon is the same as two half-gallons. Each half-gallon is the same as two quarts. One quart is the same as two pints. One pint is the same as two cups. One cup is the same as eight fluid ounces. The following table describes these relationships.

	Half-Gallons	Quarts	Pints	Cups	Ounces
Gallon	2	4	8	16	128
Half-Gallon		2	4	8	64
Quart			2	4	32
Pint				2	16
Cup					8

Table 3: English Measurement Liquid Conversions

This information will not be provided on the examinations. You are expected to know these units and relationships. You don't need to memorize all of the numbers; if you know the relationships, you can calculate the numbers.

Time – The Hour

The primary unit of time measurement in both systems (English and metric) is the hour. Hours are subdivided into 60 minutes. Each minute is the same as 60 seconds. Days have 24 hours, weeks have 7 days, and years have 365 days, except for leap years that have 366. The following table describes these relationships.

	Months	Weeks	Days	Hours	Minutes	Seconds
Year (non-leap)	12	52+	365	8,760	525,600	31,536,000
Month		Varies	Varies	Varies	Varies	Varies
Week			7	168	10,080	604,800
Day				24	1,440	86,400
Hour					60	3,600
Minute						60

Table 4: Time Conversions

SYSTEMS OF MEASUREMENT

This information will not be provided on the examinations. You are expected to know these units and relationships. You don't need to memorize all of the numbers; if you know the relationships, you can calculate the numbers.

Distance – The Mile

The primary unit of distance in the English system is the mile. Each mile consists of 1,760 yards. Each yard has three feet, and each foot has twelve inches. The following table describes these relationships.

	Yards	Feet	Inches
Mile	1,760	5,280	63,360
Yard		3	36
Foot			12

Table 5: English Measurement Distance Conversions

This information will not be provided on examinations. You are expected to know these units and relationships. You don't need to memorize all of the numbers; if you know the relationships, you can calculate the numbers.

Temperature – Degrees Fahrenheit

Temperature in the English system is most commonly expressed as degrees Fahrenheit. In 1724, Gabriel Fahrenheit used liquid mercury to measure temperature. Mercury's thermal expansion range is large and uniform. On the scale selected by Fahrenheit, the boiling point of water was 212. He adjusted the freezing point of water to 32 so that the difference would be 180. Temperatures measured on this scale are designated as degrees Fahrenheit (° F).

ENGLISH UNIT QUESTIONS

a. One half pound equals how many ounces?
b. One cubic foot equals how many cubic inches?
c. One half gallon equals how many ounces? How many pints?
d. Two hours equal how many seconds?
e. Two miles equal how many feet?

f. The boiling temperature of water on the Fahrenheit scale minus the freezing temperature yields what range?

ENGLISH UNIT ANSWERS

a. 8, because 16 oz = 1 pound
b. 1,728, because 12" high times 12" wide times 12" deep equals 1,728 inches cubed
c. 64 ounces, because 128 oz = 1 gallon, and 4 pints, because 8 pints = 1 gallon
d. 7,200, because 2 hours times 60 gives minutes; times 60 gives 7,200 seconds
e. 10,560, because 5,280 feet = one mile
f. 180 degrees, because 212 degrees minus 32 degrees yields 180 degrees

Metric Units

Metric units of measuring weight, volume, liquids, time, distance, and temperature are probably less familiar to you than English units unless you have a scientific or international background. Each type of measurement has a base unit (meters, liters, etc.) with prefixes that increase or reduce the base unit by factors of ten (kilo-, milli-, etc.). These units and prefixes will appear frequently in safety certification examinations. You will be expected to know the relationships between these units of measure.

Weight – The Gram

The primary unit of weight measure in the metric system is the gram. The weight of one gram is derived from the weight of one cubic-centimeter (one milliliter) of water at 4°C and one atmosphere of pressure.

Volume – The Cubic Meter

The most commonly used unit of larger-volume measure in the metric system is the cubic meter. One thousand liters comprise a cubic meter. Smaller volumes are typically measured in cubic-centimeters (milliliters).

Liquids – The Liter

The primary unit of liquid measure in the metric system is the liter. The liter is equal to one thousand cubic-centimeters (milliliters), and one liter of water weighs one kilogram.

Time – The Hour

Time measurement in the metric system is the same as in the English system.

Distance – The Meter

The primary unit of distance measure in the metric system is the meter.

Temperature – The Celsius Scale

The primary unit of temperature measure in the metric system is the Celsius (° C) scale. Carolus Linnaeus (1745) described a scale for which the difference between the freezing point of water (zero), and the boiling point (100), made it a Centigrade (hundred-step) scale. In 1948 use of "Centigrade" was dropped in favor of a similar scale called Celsius. Celsius temperatures are written as ° C. Although on the Celsius scale the boiling point of water at standard atmospheric pressure is 99.975 C (instead of the 100 degrees defined by the Centigrade scale) the two are virtually identical.

Prefixes – Scaling the Metric System

Instead of using completely different words to describe sub-units of weight, volume, and distance, as the English system does (for example, there are 12 "inches" in a "foot"), the metric system uses prefixes along with the base units. These prefixes, along with the base units, gram, liter, and meter, indicate how sub-units and larger units differ from the base by factors of ten. The following table shows the prefix, the multiplier, and the effect on each common unit.

Prefix	Symbol	Amount	Multiplier	No. of base grams, liters, or meters	Exponent
pico	p	One-trillionth	$\frac{1}{1,000,000,000,000}$	0.000000000001	10^{-12}
nano	n	One-billionth	$\frac{1}{1,000,000,000}$	0.000000001	10^{-9}
micro	μ (mu)	One-millionth	$\frac{1}{1,000,000}$	0.000001	10^{-6}
milli	m	One-thousandth	$\frac{1}{1,000}$	0.001	10^{-3}
centi	c	One-hundredth	$\frac{1}{100}$	0.01	10^{-2}
deci	d	One-tenth	$\frac{1}{10}$	0.1	10^{-1}
BASE		one	1	1	10^{0}
deca	da	ten	10	10	10^{1}
hecto	h	One hundred	100	100	10^{2}
kilo	k	One thousand	1,000	1,000	10^{3}
mega	M	One million	1,000,000	1,000,000	10^{6}
giga	G	One billion	1,000,000,000	1,000,000,000	10^{9}
tera	T	One trillion	1,000,000,000,000	1,000,000,000,000	10^{12}

Table 6: Metric Prefixes

SYSTEMS OF MEASUREMENT

This information will not be provided on the examinations. You are expected to know these units and relationships. You don't need to memorize all of the numbers; if you know the relationships, you can calculate the numbers.

Water – The Basis of the Metric System

The unit of a gram of water is significant in the metric system. The gram of water occupies a space equal to one cubic centimeter, abbreviated "cc." A gram of water also comprises a fluid volume equal to one milliliter (ml). This equivalence: 1 gram water = one cc = one ml, allows you to move between grams, meters, and liters. It also allows convenient calculation of other weights and volumes. Since one milliliter (thousandth of a liter) occupies one cubic centimeter, there are one thousand cubic-centimeters in a liter. Because a cubic centimeter of water weighs one gram, there are one thousand grams (a kilogram) in a liter of water. Because each cubic meter of water contains a thousand liters, the cubic meter of water weighs a thousand kilograms.

METRIC UNIT QUESTIONS

a. How many grams equal a hectogram?
b. How many liters are in a cubic meter?
c. How many cubic centimeters equal a liter?
d. How many micrometers equal a meter?
e. Is temperature in the metric system is measured in Fahrenheit, Centigrade, or Celsius?
f. How many centimeters equal a meter?
g. How many grams equal a decagram?

METRIC UNIT ANSWERS

a. 100 (because the "hecto" prefix means "times 100").
b. 1,000 (because 1,000 cubic centimeters equal a liter, and 1,000,000 cc's equal a cubic meter).
c. 1,000 (because a cubic centimeter is equal to a milliliter).
d. 1,000,000 (because the prefix "micro" means "divided by one million").

e. Celsius (Because Fahrenheit is an English measurement, and Centigrade is no longer used.).

f. 100 (because the prefix "centi" means "divided by one hundred.").

g. 10 (because prefix "deca" means "times 10").

Other Measurement Systems

The *Candidates' Handbook* for the ASP/CSP provides a chart of units of measure. The "Dynamic Units" chart looks like this:

	Force	Mass	Acceleration
English Absolute	poundal	pound	ft/sec^2
British Engineering	pound	slug (32.2 lb.)	$ft/sec.^2$
Metric Absolute	dyne	gram	cm/sec^2
SI Units	newton	kilogram	m/sec^2

Table 7: Alternate Measurement System Units

These terms are defined as follows:

Poundal: The amount of force required to accelerate a one pound mass at a rate of one foot per second.

Slug: A mass that would accelerate at the rate of one foot per second when acted on by a force of one pound. On earth, gravity exerts a force of 32.2 feet per second, so a slug would equal 32.2 pounds.

Dyne: The amount of force required to accelerate a one gram mass at a rate of one centimeter per second.

Newton: The amount of force required to accelerate a one kilogram mass at a rate of one meter per second.

SYSTEMS OF MEASUREMENT

SI Units: Standard International units that consist of the following:

Unit of Measure	S.I. Unit	Symbol
length	meter	m
mass	kilogram	kg
time	second	s
electric current	ampere	A
temperature	kelvin	K
amount	mole	mol
brightness	candela	cd

Table 8: SI Unit Conventions

Absolute Temperatures

Kelvin and Rankin temperature scales are used for scientific work. These two scales are used as "absolute" temperature scales for chemistry equations.

Kelvin differs from Celsius in that there are no below-zero Kelvin temperatures. The temperature of 0K represents absolute zero, or the absence of any heat. The size of a Kelvin degree is the same as the size of a Celsius degree. To convert Celsius to Kelvin, add 273 to the Celsius temperature. To convert from Kelvin to Celsius, subtract 273 from the Kelvin temperature. The *Candidates' Handbook* gives this formula to define the relationship between Kelvin and Celsius systems:

$t_{^0K} = t_{^0C} + 273$.

The Rankin scale differs from Fahrenheit in that there are no below-zero Rankin temperatures. The temperature 0R, like 0K, represents absolute zero. The size of a Rankin degree is the same as the size of a Fahrenheit degree. To convert Fahrenheit to Rankin, add 460 to the Fahrenheit temperature. To convert from Rankin to Fahrenheit, subtract 460 from the Rankin temperature. The *Candidates' Handbook* gives this formula to define the relationship between Fahrenheit and Rankin systems:

$t_{^0R} = t_{^0F} + 460$.

Absolute Pressures

In working chemistry equations, absolute pressures must be used. Common "gauge" pressure (psig or pounds-per-square-inch-gauge) is calibrated to read zero at ambient pressure. This is inadequate for chemistry work because when the gauge shows zero, the actual air pressure is 14.7 pounds per square inch. The reason for this is that the surface of the earth is at the bottom of a sea of air, the weight of which exerts increasingly greater force as you approach sea level. This force (14.7 pounds per square inch) is considered absolute (psia or pounds-per-square-inch-absolute). This psia force must be used for chemistry equations, rather than the gauge pressure. To convert psig to psia, add 14.7 to the psig pressure. To convert from psia to psig, subtract 14.7 from the psia pressure.

Other pressures that can be used as absolute pressure for chemistry equations are given in the "Standards" section of the equation sheets in the *Candidates' Handbook*:

1 atm =   14.7 psi (psia)

760 mm Hg (millimeters of mercury)

29.92 in Hg (inches of mercury)

33.90 ft H_2O

760 torr (A torr is equal to $\frac{1}{760}$ mm of mercury (Hg).)

101.3 kilopascal (A kilopascal is the pressure exerted by a 10g mass resting on a $1 cm^2$ area.)

OTHER MEASUREMENT SYSTEMS QUESTIONS

a. The definition "The amount of force required to accelerate a one gram mass by a rate of one centimeter per second" is the description of what term?

b. The Standard International (SI) unit for temperature is what?

c. Which of the following scales are absolute? Kelvin, Fahrenheit, Celsius, Rankin.

d. Convert 186 degrees Fahrenheit to degrees Rankin.

e. Convert 470 degrees Kelvin to degrees Celsius.

f. Which of the following terms is not an absolute pressure: 1 atmosphere, 760 mm Hg, 14.7 psig?

SYSTEMS OF MEASUREMENT

OTHER MEASUREMENT SYSTEMS ANSWERS

a. Dyne (these descriptions must be memorized).

b. The Kelvin degree (these descriptions must be memorized).

c. Kelvin and Rankin (because Fahrenheit and Celsius are English & Metric scales, respectively. (Centigrade is no longer used.).

d. 646 R (because degrees F + 460 = Degrees R).

e. 197 (because degrees C + 273 = degrees K. To go from K to C, subtract 273).

f. 14.7 psig (because 0 psig is equal to 14.7 psia).

Conversions

You will frequently be required to convert from English to Metric units or vice versa on safety certification examinations. Some scientific calculators have conversions built into them and some do not. If your calculator lacks conversions, the *Candidates' Handbook* offers the following conversion factors:

Length:	1 in = 2.54 cm
	1 ft = 30.48 cm
	1 micron = 10^{-4} cm
	1 Angstrom = 10^{-8} cm
Volume:	1 liter = 1.06 qt
	1 liter = 61.02 in^3
	1 liter = 0.03531 ft^3
Mass:	1 kg = 2.2 lb
	1 lb = 454 gm
Temperature:	$t_{°C} = \dfrac{(t_{°F} - 32)}{1.8}$
	$t_K = t_C + 273$
	$t_R = t_F + 460$

Table 9: Conversions Given by *Candidates' Handbook*

CHAPTER 4

Some candidates prefer to memorize "magic numbers" for conversions. These allow direct conversion from English to metric units. The table below gives common conversion factors:

	When you know:	Multiply by:	To get:
Length	Inches (in.)	2.54	Centimeters (cm)
	Feet (ft.)	30.48	Centimeters (cm)
	Yards (yd.)	0.91	Meters (m)
	Miles (mi.)	1.60	Kilometers (km)
	Millimeters (mm)	0.04	Inches (in.)
	Centimeters (cm)	0.39	Inches (in.)
	Meters (m)	3.28	Feet (ft.)
	Kilometers (km)	0.62	Miles (mi.)
Area	Square Inches (in^2)	6.45	Square Centimeters (cm^2)
	Square Feet (ft^2)	0.09	Square Meters (m^2)
	Square Yards (yd^2)	0.83	Square Meters (m^2)
	Square Miles (mi^2)	2.58	Square Kilometers (km^2)
	Square Centimeters (cm^2)	0.16	Square Inches (in^2)
	Square Meters (m^2)	10.76	Square Feet (ft^2)
	Square Meters (m^2)	1.19	Square Yards (yd^2)
	Square Kilometers (km^2)	0.38	Square Miles (mi^2)
Mass	Ounces (oz)	28.34	Grams (g)
	Pounds (lb)	0.45	Kilograms (kg)
	Short Tons (2,000 lb)	0.90	Tonnes (t)
	Grams (g)	0.035	Ounces (oz)
	Kilograms (kg)	2.20	Pounds (lb)
	Tonnes (1,000 kg)	1.10	Short Tons

SYSTEMS OF MEASUREMENT

Volume	Fluid Ounces (fl oz)	29.57	Milliliters (ml)
	Pints (pt)	0.47	Liters (l)
	Quarts (qt)	0.95	Liters (l)
	Gallons (gal)	3.78	Liters (l)
	Imp. Gallons	4.55	Liters (l)
	Milliliters (ml)	0.03	Fluid Ounces (fl oz)
	Liters (l)	2.11	Pints (pt)
	Liters (l)	1.06	Quarts (qt)
	Liters (l)	0.26	Gallons (gal)
	Liters (l)	0.22	Imperial Gallons
Temperature	Fahrenheit °F	$(F - 32) \cdot \frac{5}{9}$	Celsius °C
	Celsius °C	$\left(C \cdot \frac{9}{5}\right) + 32$	Fahrenheit °F
Water Pressure	Vertical column of water in feet (multiply by specific gravity if not water)	0.433	Pounds per square inch (psi)

Table 10: Conversion Factor Magic Numbers

Dimensional Analysis

Dimensional analysis is a tool for converting between units. The problems you must solve on the safety certification examinations are unlikely to be exactly in the units provided by your calculator or in the units provided on the equation sheets. This means that you'll need to go through several steps in order to arrive at the desired unit. The sequence is as follows:

First, write the term to be converted (both number and unit).

Second, write a fraction, equal to one, that has the same unit in the opposite position (if the unit to be converted from step one is in the numerator, put the same unit in the denominator of the fraction).

Third, multiply the terms and cancel the units.

Fourth, round the units as needed to achieve accuracy.

EXAMPLE

To convert 20 miles to feet:

First write the term to be converted: 20 miles

Second, write a fraction equal to one that puts miles in the denominator: $\frac{5,280 \ feet}{1 \ mile}$

Third, multiply terms and cancel units: $\frac{20 \ miles}{1} \cdot \frac{5,280 \ feet}{1 \ mile}$. The miles cancel out leaving only feet: $\frac{20 \cdot 5,280 \cdot feet}{1} = 105,600$ feet.

Fourth, round the 105,600 if needed. Rounding is not needed in this case, because all units are whole numbers.

This sequence can be performed over several steps to provide multiple unit conversions. The steps are the same, although the second step is usually performed using a continuous division line as follows:

SYSTEMS OF MEASUREMENT

EXAMPLE

Convert 3 hours to seconds:

$$\frac{3 \text{ hours}}{} \cdot \frac{60 \text{ minutes}}{1 \text{ hour}} \cdot \frac{60 \text{ seconds}}{1 \text{ minute}}$$

Once the division line is set up, cancel units:

$$\frac{3 \text{ \sout{hours}}}{} \cdot \frac{60 \text{ \sout{minutes}}}{1 \text{ \sout{hour}}} \cdot \frac{60 \text{ seconds}}{1 \text{ \sout{minute}}}$$

And multiply to achieve $\frac{3 \cdot 60 \cdot 60 \cdot sec}{1} = 10{,}800$ seconds. Then round if needed. Rounding is not needed in this case.

The technique also works when you have units in both the numerator and denominator.

EXAMPLE

Convert 60 mph (miles per hour) to fps (feet per second)

First, set up the number line to convert the miles to feet:

$$\frac{60 \text{ miles}}{1 \text{ hour}} \cdot \frac{5{,}280 \text{ feet}}{1 \text{ mile}}$$

Note that the 5,280 feet = 1 mile is a conversion factor that you must memorize.

Next add the conversion from hours to seconds to the same number line. Since the hour unit appears in the denominator of the original fraction, the hour unit of the conversion factor will appear in the numerator:

$$\frac{60 \text{ miles}}{1 \text{ hour}} \cdot \frac{5{,}280 \text{ feet}}{1 \text{ mile}} \cdot \frac{1 \text{ hour}}{60 \text{ minutes}} \cdot \frac{1 \text{ minute}}{60 \text{ seconds}}$$

59

Next, cancel units that appear in both the numerator and denominator:

60 ~~miles~~	5,280 feet	1 ~~hour~~	1 ~~minute~~
1 ~~hour~~	1 ~~mile~~	60 ~~minutes~~	60 seconds

Next, multiply numbers to achieve: $\dfrac{60 \cdot 5{,}280 \cdot feet}{60 \cdot 60 \cdot sec} = 88$ feet per second. Then round if needed. Rounding is not needed in this case.

The same dimensional analysis technique can be used to convert units between units in the English and metric systems.

EXAMPLE

Convert 40 mph (miles per hour) to meters per second:

40 miles	5280 feet	30.48 cm	1 meter	1 hour	1 minute
1 hour	1 mile	1 foot	100 cm	60 minutes	60 seconds

Next, cancel units and perform the actual multiplication. Note that in setting up the above, you have to know that 5,280 feet = 1 mile and that 100 cm = 1 meter. The 30.48cm = 1 foot conversion is given in the "Unit Conversions" section of the *Candidates' Handbook* formula sheet.

40 ~~miles~~	5280 ~~feet~~	30.48 ~~cm~~	1 meter	1 ~~hour~~	1 ~~minute~~
1 ~~hour~~	1 ~~mile~~	1 ~~foot~~	100 ~~cm~~	60 ~~minutes~~	60 seconds

Leaving: $\dfrac{40 \cdot 5280 \cdot 30.4 \cdot meters}{100 \cdot 60 \cdot 60 \cdot sec}$, or 17.8816 meters per second

SYSTEMS OF MEASUREMENT

CONVERSIONS AND DIMENSIONAL ANALYSIS QUESTIONS

a. Four inches equals how many centimeters?
b. Two liters equals how many cubic inches?
c. 5,280 yards equal how many meters?
d. 287.0 decagrams equal how many kilograms?
e. Convert 40 kilometers per hour to feet per second.
f. Convert 2,850 pounds per day to grams per minute.
g. A vapor pressure of 28 grams per square meter exists on a diaphragm. Convert to psi.

CONVERSIONS AND DIMENSIONAL ANALYSIS ANSWERS

a. 10.16, because the *Candidates' Handbook* gives the conversion factor, 1 in = 2.54 cm. If one inch is 2.54 cm, then 4 inches times 2.54 yields 10.16 cm.

b. 122.04, because the *Candidates' Handbook* gives the conversion factor, 1 liter = 61.02 in^3. If one liter is 61.02 then two liters times 61.02 gives 122.04 cubic inches.

c. 4,828 meters. To arrive at this answer, first set up a dimensional analysis line by multiplying by fractions equal to one which move from the unit you want to convert from (yards) toward the unit you want to convert to (meters). Note that the 1 inch = 2.54 cm conversion is given in the *Candidates' Handbook*.

$$\frac{5{,}280 \text{ yards}}{} \cdot \frac{36 \text{ inches}}{1 \text{ yard}} \cdot \frac{2.54 \text{ cm}}{1 \text{ inch}} \cdot \frac{1 \text{ meter}}{100 \text{ cm}}$$

Next, cancel terms and perform the multiplication:

$$\frac{5{,}280 \cancel{\text{ yards}}}{} \cdot \frac{36 \cancel{\text{ inches}}}{1 \cancel{\text{ yard}}} \cdot \frac{2.54 \cancel{\text{ cm}}}{1 \cancel{\text{ inch}}} \cdot \frac{1 \text{ meter}}{100 \cancel{\text{ cm}}}$$

$$= \frac{5{,}280 \bullet 36 \bullet 2.54 \bullet meters}{100} = 4{,}828.032 \; meters$$

CHAPTER 4

Last, round to a whole number, because the original distance (5,280 yards) was a whole number: 4,828 meters.

d. 2.9 kilograms. To arrive at this answer, first set up a dimensional analysis line, multiplying by fractions equal to one which move from the unit you want to convert from (decagrams) toward the unit you want to convert to (kilograms). No conversions between metric and English systems are needed in this sequence, but you must know the meaning of the prefixes.

287.0 decagrams	1.0 hectogram	1.0 kilograms
	10.0 decagram	10.0 hectogram

Next, cancel out terms and perform the multiplication:

287.0 ~~decagrams~~	1.0 ~~hectogram~~	1.0 kilograms
	10.0 ~~decagrams~~	10.0 ~~hectograms~~

This equals $\frac{287.0 \cdot kilograms}{100}$, or 2.87 kilograms. Last, round to one decimal place, because numbers of that accuracy were the least precise units used in the conversion sequence: 2.9 kilograms.

e. 36 feet per second. To arrive at this answer, first set up a dimensional analysis line by multiplying by fractions equal to one which move from the units you want to convert from (kilometers and hours) toward the units you want to convert to (feet and seconds).

40 kilometers	1,000 meters	100 centimeters	1 foot	1 hour	1 minute
1 hour	1 kilometer	1 meter	30.48 centimeters	60 minutes	60 seconds

Next, cancel terms and perform the multiplication. Note that the 30.48 cm = 1 foot conversion is given in the *Candidates' Handbook*.

SYSTEMS OF MEASUREMENT

40 ~~kilometers~~	1,000 ~~meters~~	100 ~~centimeters~~	1 foot	1 ~~hour~~	1 minute
1 ~~hour~~	1 ~~kilometer~~	1 ~~meter~~	30.48 ~~centimeters~~	60 ~~minutes~~	60 seconds

This yields an answer of 36.453775 feet per second. Since the least accurate number used in the conversion sequence was the whole number of 40 kilometers given in the original question, the answer rounds to 36 feet per second.

f. 900 grams per minute. To arrive at this answer, first set up a dimensional analysis line by multiplying by fractions equal to one which move from the units you want to convert from (pounds and days) toward the units you want to convert to (grams and minutes).

2,850 pounds	1 kilogram	1,000 grams	1 day	1 hour
1 day	2.2 pounds	1 kilogram	24 hours	60 minutes

Next, cancel out terms and perform the multiplication. Note that the 2.2 pounds = 1 kg conversion is given in the *Candidates' Handbook*.

2,850 ~~pounds~~	1 ~~kilogram~~	1,000 grams	1 ~~day~~	1 ~~hour~~
1 ~~day~~	2.2 ~~pounds~~	1 ~~kilogram~~	24 ~~hours~~	60 minutes

This yields an answer of approximately 899.621 grams per minute. Since the least accurate number used in the conversion sequence was a whole number, this rounds to 900 grams per minute.

g. $3.97 \bullet 10^{-5}$ psi, because:

28 ~~grams~~	1 ~~kilogram~~	2.2 pounds	$(1 \text{ ~~meter~~})^2$	$(2.54 \text{ ~~cm~~})^2$
$(1 \text{ ~~meter~~})^2$	1,000 ~~grams~~	1 ~~kilogram~~	$(100 \text{ ~~cm~~})^2$	$(1 \text{ inch})^2$

Which becomes the following after the squared values are applied to the numbers:

$$\frac{28 \text{ grams}}{(1 \text{ meter})^2} \left| \frac{1 \text{ kilogram}}{1{,}000 \text{ grams}} \right| \frac{2.2 \text{ pounds}}{1 \text{ kilogram}} \left| \frac{(1 \text{ meter})^2}{10{,}000 \text{ cm}^2} \right| \frac{6.45 \text{ cm}^2}{(1 \text{ inch})^2}$$

Multiplying, we have: $\dfrac{397.42\#}{10{,}000{,}000 in^2}$ which is the same as $3.97 \bullet 10^{-5}$ pounds per one square inch (psi)

SYSTEMS OF MEASUREMENT CHALLENGE EXAM

a. 56 ounces equals how many pounds?
b. 2,592 cubic inches equals how many cubic feet?
c. 2.5 quarts yield how many pints?
d. 0.5 gallons yields how many ounces?
e. One half mile is equal to how many feet?
f. One nanogram equals how many picograms?
g. 1.5 cubic meters equal how many liters?
h. 1,250 cc's equal how many liters?
i. How many decimeters in a hectometer?
j. 2,467 cc's of water weigh what in kilograms?
k. The amount of force required to accelerate a one kilogram mass by a rate of one meter per second is the definition of what unit?
l. 776 degrees Rankin is what temperature in Fahrenheit?
m. 200°C is what temperature in absolute units?
n. 760 Torr is an absolute pressure, True or False?
o. 14.7 psig is an absolute pressure, True or False?
p. Convert 276.00 miles per hour to kilometers per second.
q. A column of liquid exerts a force on the bottom of a tank of 50 psi. Convert this to kg/meter^2.

SYSTEMS OF MEASUREMENT CHALLENGE EXAM ANSWERS

a. 3.5, because $16oz = 1lb$, so $\dfrac{56oz \bullet 1lb}{16oz} = 3.5lb$

b. 1.5, because $1,723in^3 = 1ft^3$, so $\dfrac{2,592in^3 \bullet 1ft^3}{1,723in^3} = 1.5ft^3$

c. 5, because $2\,pints = 1qt$, so $\dfrac{2.5qt \bullet 2\,pints}{1qt} = 5\,pints$

d. 64, because $\dfrac{1}{2}gal. = 2qts = 4\,pints = 8cups = 64oz.$

e. 2,640, because $5,280ft = 1mi$, so $\dfrac{0.5mi \bullet 5,280ft}{1mi} = 2,460ft$

f. 1,000, because $\dfrac{1ng \bullet 1,000\,pg}{1ng} = 1,000\,picograms$

g. 1,500, because $1,000l = 1m^3$, so $\dfrac{1.5m^3 \bullet 1,000l}{1m^3} = 1,500\,liters$

h. 1.25, because $1,000cc = 1liter$, so $\dfrac{1,250cc \bullet 1liter}{1,000cc} = 1.25\,liters$

i. 1,000, because $10\,decimeters = 1meter$ & $100\,meters = 1hectometer$, so

$\dfrac{10\,decimeters \bullet 100\,meters}{1meter \bullet 1hectometer} = \dfrac{1,000\,decimeters}{1hectometer}$

j. 2.467, because $1,000cc(water) = 1kg$, so $\dfrac{2,467cc(water) \bullet 1kg}{1,000cc(water)} = 2.467kg$

k. The Newton (by definition)

l. 316 degrees Fahrenheit, because 776R-460=316F

m. 473 Kelvin, because 200C+273=473K

n. True   (given in *Candidates' Handbook*)

65

o. False   (psi*a* – not psig is absolute)

p. 0.12 kilometers per second, because:

$$\frac{276.00 \text{ miles}}{1.00 \text{ hr}} \cdot \frac{5{,}280.00 \text{ ft.}}{1.00 \text{ mile}} \cdot \frac{30.48 \text{ cm}}{1.00 \text{ ft.}} \cdot \frac{1.00 \text{ meters}}{100.00 \text{ cm}} \cdot \frac{1.00 \text{ km}}{1{,}000.00 \text{ meters}} \cdot \frac{1.00 \text{ hr}}{60.00 \text{ min}} \cdot \frac{1.00 \text{ min}}{60.00 \text{ sec}}$$

This is equal to $\dfrac{276.00 \bullet 5{,}280.00 \bullet 30.48 \bullet km}{100.00 \bullet 1{,}000.00 \bullet 60 \bullet 60 \bullet \sec} = 0.123383$, which rounds to 0.12

q. $\dfrac{34{,}751 kg}{m^2}$, because:

$$\frac{50 \text{ pounds}}{(1 \text{ inch})^2} \cdot \frac{1 \text{ kg.}}{2.2 \text{ pounds}} \cdot \frac{(1 \text{ inch})^2}{(2.54 \text{ cm})^2} \cdot \frac{(100 \text{ cm})^2}{(1 \text{ meter})^2}$$

Which converts to:

$$\frac{50 \text{ pounds}}{(1 \text{ inch})^2} \cdot \frac{1 \text{ kg.}}{2.2 \text{ pounds}} \cdot \frac{(1 \text{ inch})^2}{6.45 \text{ cm}^2} \cdot \frac{10{,}000 \text{ cm}^2}{(1 \text{ meter})^2}$$

This is the same as $\dfrac{500{,}000 kg.}{14.19 m^2}$, or 34,751 kg per square meter.

CHAPTER FIVE

Scientific and Engineering Notation

> **Assumptions:** You have comprehended the material in previous chapters, and you understand exponents. You have a scientific calculator and knowledge of how to use it.
>
> **Application:** Required for ASP, CSP, OHST, and CHST examinations.
>
> **Discussion:** Scientific and Engineering notation presumed by the BCSP to be common knowledge material that candidates are expected to know. Scientific notation will appear on all examinations. Engineering is less common. If you feel your comprehension of scientific and engineering notation are good, go to the end of this chapter and try the challenge exam. If you can work all of the problems correctly, skip this chapter.

Scientific Notation

Scientific notation is a way of conveniently expressing very large or very small numbers. For example, it is easier to write $1 \bullet 10^9$ than to write 1,000,000,000. Scientific notation consists of a number called the *base* multiplied by ten raised to a whole-number exponent. Expressing a number in scientific notation is also more likely to be accurate than expressing it using lots of zeros, since counting the zeros on a calculator's display (with no commas shown) is a known source of mistakes.

In scientific notation, the number to the left of the decimal point in the base must be greater than or equal to one and less than ten. Any number of digits is acceptable to the right of the decimal point. Note that there must be one and only one non-zero digit to the left of the decimal in a base.

For example, 9.99 is an acceptable base, but 99.9 is not.

Once you have established the base, count places from the decimal point of the base to the decimal point of the original number. The number of places you moved the decimal point right or left will determine the exponent of 10 that the base will be multiplied by.

EXAMPLE

In the number 1,004,000, the base will become 1.004. Since you moved the decimal point six places from the original number to the base, the base of ten will be multiplied to the power of six. Because you moved the decimal point to the left (from the original number to the new base), the exponent will be positive. This means that 1,004,000 expressed in scientific notation is $1.004 \bullet 10^6$.

If you move the decimal point of the original number to the left to arrive at the base, the exponent of ten will be positive. If you move the decimal point to the right, the exponent will be negative.

EXAMPLE

The base of 0.0002986 is 2.986. Because you moved the decimal point four places to the right to obtain the base, the exponent of ten will be a negative four. This means that 0.0002986 expressed in scientific notation is $2.986 \bullet 10^{-4}$

Another example: 1,234,567 will yield a base of 1.234567. Because you moved the decimal point six places to the left to obtain the base, the exponent of ten will be a positive six. This means that 1,234,567 expressed in scientific notation is $1.234567 \bullet 10^6$.

Some calculators have keys that convert from decimal form to scientific notation and back. Others have a shorthand entry key for numbers already in scientific notation. This key is usually marked "E," "EE," or "EXP." Usually, the base is keyed into the calculator first, the EE key is pressed, and the exponent is entered with its sign (+ or –). These keystrokes vary from calculator to calculator, so consult your calculator's operating manual for instructions.

To convert from scientific notation back to decimal form, use your calculator to display the decimal form of the number (if your calculator has this conversion). If your calculator will not do the conversion, move the decimal point the number of places shown by the ten's exponent. If the exponent is positive, move the decimal point to the right. If the exponent is negative, move the decimal point to the left.

EXAMPLE

To convert $2.004 \bullet 10^3$ back to decimal form, start with the base (2.004), and move the decimal point three places to the right: 2,004.

Another example: To convert $2.004 \bullet 10^{-4}$ to decimal form, start with the base (2.004) and move the decimal point four places to the left: 0.0002004.

SCIENTIFIC AND ENGINEERING NOTATION

To multiply in scientific notation, bases multiply, but exponents add.

To multiply numbers in scientific notation, it is easier to let the calculator do the math. If you wish to do it manually, note that bases are multiplied, but exponents added.

EXAMPLE

To multiply $2.04 \bullet 10^4$ by $3.007 \bullet 10^{-2}$, first multiply the bases $(2.04 \bullet 3.007) = 6.13428$. Then add the exponents: $4 + (-2) = 2$. Finally, put the expression back together again: $6.13428 \bullet 10^2$, or 613.428.

To divide in scientific notation, bases divide, but exponents subtract.

To divide numbers in scientific notation, it is easier to let the calculator do the math. If you wish to do it manually, note that bases are divided, but exponents subtracted.

EXAMPLE

$\dfrac{4.78 \bullet 10^6}{2.47 \bullet 10^3}$ can be simplified by first dividing the bases: $\dfrac{4.78}{2.47} = 1.9352226$. Then subtract the exponent in the denominator from the exponent in the numerator: $6 - 3 = 3$. Finally, put the expression back together again: $1.9352226 \bullet 10^3$, or $1{,}935.2226$.

Addition and subtraction of numbers in scientific notation is best done by the calculator.

SCIENTIFIC NOTATION QUESTIONS

a. Is $28.6 \bullet 10^4$ in scientific notation?

b. Express 1,234,005,000 in scientific notation.

c. Express 0.001000285001 in scientific notation.

d. Convert $5.28 \bullet 10^{-4}$ to decimal form.

e. Convert $1.2345 \bullet 10^7$ to decimal form.

CHAPTER 5

f. $(1.385 \bullet 10^{-5}) \bullet (4.2856 \bullet 10^{13}) = ?$

g. $\dfrac{2.85 \bullet 10^7}{1.68 \bullet 10^9} = ?$

h. $\dfrac{1.285 \bullet 10^8}{3.721 \bullet 10^7} = ?$

SCIENTIFIC NOTATION ANSWERS

a. No, because the base must have only one digit between one and ten (not including ten) to the left of the decimal point.

b. $1.234005 \bullet 10^9$, because the base must be between 1 and 10 (not including ten), and the decimal point moved nine places to the left between the original number and the scientific notation base. The fact that the decimal moved to the left makes the exponent (nine) positive.

c. $1.000285001 \bullet 10^{-3}$, because the base must be between 1 and 10 (not including ten), and the decimal point moved three places to the right between the original number and the scientific notation base. The fact that the decimal moved to the right makes the exponent (three) negative.

d. 0.000528, because the decimal point moves four places to the left since the exponent is negative.

e. 12,345,000, because the decimal point moves seven places to the right since the exponent is positive.

f. $5.935556 \bullet 10^8$, because the bases are multiplied $(1.385 \bullet 4.2856 = 5.935556)$, and the exponents are added $(-5 + 13 = 8)$.

g. $1.6964285 \bullet 10^{-2}$, because the bases are divided $\left(\dfrac{2.85}{1.68} = 1.6964285\right)$, and the exponents are subtracted: $(7 - 9 = -2)$.

SCIENTIFIC AND ENGINEERING NOTATION

h. 3.453372, because the bases are divided: $\frac{1.285}{3.721} = 0.3453372$, and the exponents are subtracted: $8 - 7 = 1$. This leaves $0.3453372 \bullet 10^1$. This is not acceptable scientific notation because the base has no non-zero number to the left of the decimal. To convert to scientific notation, we must move the decimal one place to the right, and reduce the exponent by one. This makes the expression $3.453372 \bullet 10^0$. Since any number to the zero power is one, the base is the final answer: 3.453372.

Engineering Notation

Engineering notation works the same as scientific notation except that the base may have up to three digits to the left of the decimal point and all exponents must be multiples of three. Since the base may have up to three digits, any number between 1 and 999 (inclusive) is acceptable. Exponents are always positive or negative multiples of three.

EXAMPLE

To convert 0.0002896 to engineering notation, first move the decimal to the right in increments of three. The first jump gives us $0.2896 \bullet 10^{-3}$. This is not acceptable because it doesn't produce a non-zero base. The next jump of three decimal places to the right produces: $289.6 \bullet 10^{-6}$. This is an acceptable base (it's between 1 and 999), and the exponent is an even multiple of three. The number is now in engineering notation.

Another example: To convert 10,289 to engineering notation, first move the decimal to the left in increments of three. The result: $10.289 \bullet 10^3$ meets the criteria for bases and exponents and is now in engineering notation.

Engineering notation may or may not be used on the safety certification examinations, but you should be familiar with it just in case. Some calculators also have a key that toggles between scientific and engineering notation. Don't mistake one for the other. Multiplication and division in engineering notation work exactly the same as in scientific notation.

CHAPTER 5

ENGINEERING NOTATION QUESTIONS

a. Convert 1,000,028,000,000 to engineering notation.

b. Convert 0.000000382017 to engineering notation.

c. Convert $4.287 \bullet 10^5$ to engineering notation.

ENGINEERING NOTATION ANSWERS

a. $1.000028 \bullet 10^{12}$. No acceptable base is found by moving the decimal point three, six, or nine places to the left; the only acceptable base is reached when the decimal point is moved 12 places.

b. $382.017 \bullet 10^{-9}$. No acceptable base is found by moving the decimal point three or six places to the right; the only acceptable base occurs when the decimal point is moved nine places to the right.

c. $428.7 \bullet 10^3$, because the original scientific notation converts to decimal form as 428,700. To switch to engineering notation, move the decimal three places to the left. The base 428.7 is acceptable.

SCIENTIFIC AND ENGINEERING NOTATION CHALLENGE EXAM

a. Express 12,000,287.42 in scientific notation.

b. Express 0.002004006 in scientific notation.

c. Express $9.2867 \bullet 10^{-5}$ in decimal form.

d. Express $9.999 \bullet 10^6$ in decimal form.

e. Express $(2.87 \bullet 10^3) \bullet (4.25 \bullet 10^4) \bullet (1.0 \bullet 10^{-5})$ in scientific notation.

f. Express $\dfrac{4.5 \bullet 10^4}{9.0 \bullet 10^6}$ in scientific notation.

g. Express 123,876.54 in engineering notation.

h. Express 0.00003528 in engineering notation.

i. Express $\left(\dfrac{1.285 \bullet 10^4}{45.287 \bullet 10^3}\right) \bullet 2.78 \bullet 10^0$ in scientific notation.

SCIENTIFIC AND ENGINEERING NOTATION

SCIENTIFIC AND ENGINEERING NOTATION CHALLENGE EXAM ANSWERS

a. Approximately $1.2 \bullet 10^7$, because the only acceptable base is equal to or greater than one and less than ten, and the decimal point moved seven places to the left from the original number to the new base (making the exponent positive).

b. Approximately $2 \bullet 10^{-3}$, because the only acceptable base is equal to or greater than one and less than ten, and the decimal point moved three places to the right from the original number to the new base (making the exponent negative).

c. Approximately 0.000093, because the decimal point moved five places to the left (from the scientific notation's base to the decimal form).

d. 9,999,000, because the decimal point moved six places to the right (from the scientific notation's base to the decimal form).

e. Approximately $1.22 \bullet 10^3$. First, multiply the bases: $(2.87 \bullet 4.25 \bullet 1.0 = 12.1975)$. Then, add the exponents: $(3+4+(-5)) = (7-5) = 2$. The result is $12.1975 \bullet 10^2$. This is not in scientific notation because the base is not less than ten. To write the expression in scientific notation, move the decimal point one place to the left. This makes the new base 1.21975. Since the decimal point was moved to the left, the exponent increases by one, and the final expression is $1.21975 \bullet 10^3$, or approximately $1.22 \bullet 10^3$.

f. $5 \bullet 10^{-3}$. First divide the bases: $\dfrac{4.5}{9.0} = 0.5$. Then subtract the exponents: 4-6 = -2. The resulting expression is $0.5 \bullet 10^{-2}$. This is not in scientific notation because the base is not greater than or equal to one. To write the expression to scientific notation, move the decimal point one place to the right. This makes the new base 5.0. Since the decimal point was moved to the right, the exponent decreases by one, and the final expression is $5.0 \bullet 10^{-3}$.

g. Approximately $123.9 \bullet 10^3$. Move the decimal point three places to the left to achieve an acceptable base (between 1 and 999, inclusive), 123.87654. The final expression is $123.87654 \bullet 10^3$.

h. $35.28 \bullet 10^{-6}$. Move the decimal point six places to the right to achieve an acceptable base, 35.28. The final expression is $35.28 \bullet 10^{-6}$.

i. Approximately $7.89 \bullet 10^{-1}$. Work inside the parentheses first. Divide the bases: $\frac{1.285}{45.287} = 0.0283745$. Then, subtract the exponents: 4-3 = 1. The result of these operations is $0.0283745 \bullet 10^1$. Then multiply by the next expression, $2.78 \bullet 10^0$: Multiply the bases: $0.0283745 \bullet 2.78 = 0.0788811$, and add the exponents: 1+0 = 1. The result of these operations is $0.0788811 \bullet 10^1$. This is not in scientific notation, because the base must have a non-zero digit to the left of the decimal point. Move the decimal point two places to the right to achieve a base of 7.88811. Because you moved the decimal point two places to the right, you must subtract two from the exponent (one). The final expression is $7.88811 \bullet 10^{-1}$.

CHAPTER SIX

Algebraic Properties and Simple Equations

> **Assumptions:** You have sufficient knowledge to take and pass the challenge exam of all previous chapters. You have a scientific calculator and basic knowledge of how to use it.
>
> **Application:** Required for ASP, CSP, OHST, and CHST examinations.
>
> **Discussion:** An understanding of basic algebra is essential to practical mathematics. You will be required to solve equations by algebraic manipulation in both Safety Certification examinations and in real life. If you feel your comprehension of basic algebra is good, go to the challenge exam at the end of this chapter. If you can work all of the problems correctly, you may skip this chapter.

Variables

In algebra, a letter called a *variable* is often used as a substitute for an unknown number. In this book, lower case letters (x, y, z, a, b, c, etc.) will be used as variables. An algebraic expression contains at least one variable.

Commutative Properties

You need not know the term "commutative," but you must know how to apply commutative principles. The commutative property applies to addition and multiplication.

The commutative property for addition says:

For any real numbers a and b, a+b is equal to b+a.

This means that you can change the order of addition without affecting the sum.

EXAMPLE

$239 + 27$ is the same as $27 + 239$. In both cases, the sum (266) is the same.

The property works with more than two numbers as well. $4+5+7+9$ yields the same sum (25), no matter how you mix up the order of addition.

The commutative property for multiplication says:

For any real numbers a and b, ab is equal to ba.

This means that you can change the order of multiplication without affecting the product.

> **EXAMPLE**
>
> $239 \bullet 27$ is the same as $27 \bullet 239$. In both cases, the product (6,453) is the same.
>
> The property works with more than two numbers as well. $2 \bullet 3 \bullet 4 \bullet 5$ yields the same product (120), no matter how you mix up the order of multiplication.

Associative Properties

You need not know the term "associative," but you must know how to apply the associative principles. Like the commutative property, the associative property applies to addition and multiplication.

The associative property for addition says:

For any real numbers a, b, and c, a+(b+c) is equal to (a+b)+c.

This means that you can change the grouping of sequences of addition without affecting the sum.

> **EXAMPLE**
>
> $4+(5+6)=4+(11)=15$. Also, $(4+5)+6=(9)+6=15$. In both cases, the sum (15) is the same. Which pair of numbers was added first makes no difference. This property is valid even if more than three numbers are added.

The associative property for multiplication says:

For any real numbers a, b, and c, a(bc) is equal to (ab)c.

This means that you can change the grouping of sequences in multiplication without affecting the product.

ALGEBRAIC PROPERTIES AND SIMPLE EQUATIONS

EXAMPLE

$2 \bullet (3 \bullet 4) = 2 \bullet (12) = 24$. Also, $(2 \bullet 3) \bullet 4 = (6) \bullet 4 = 24$. In both cases, the product (24) is the same. Which pair of numbers is multiplied first makes no difference. This property is valid even if more than three numbers are multiplied.

The associative property does not apply to mixed expressions.

You cannot regroup between addition and multiplication.

EXAMPLE

$2 \bullet (3+4)$ is NOT the same as $(2 \bullet 3)+4$. In the first case, the result is 14; in the second it's 10. To see how to deal with mixed expressions, you must use the:

Distributive Property

This is sometimes called the "distributive property of multiplication over addition." This means that if a sum or difference inside parentheses is multiplied by a number outside the parentheses, each number inside the parentheses is multiplied by the number outside the parentheses.

For any real numbers a, b, and c, a(b+c) is equal to ab+ac.

EXAMPLE

$2 \bullet (3+4)$ is the same as $(2 \bullet 3)+(2 \bullet 4)$. In both cases, the simplified expression is equal to 14. You can show this by working both out: $2 \bullet (3+4) = 2 \bullet (7) = 14$, and $(2 \bullet 3)+(2 \bullet 4) = (6)+(8) = 14$.

Factoring

Factoring is undoing what's been done by the distributive property.
Since $5 \cdot (a+b) = 5a + 5b$, you can also say that $7x + 7y = 7 \cdot (x+y)$. This is useful sometimes when trying to isolate variables in an equation.

Collecting Like Terms

Expressions like 4x+4y+4z, 12a, or 10b are composed of numbers and variables called terms. Where variables are *identical*, (such as in 4y and 6y), the expressions are called "like terms." Terms such as $4y^2$ and $4y$ are not like terms because the variables are not identical (one y-variable is squared and the other is to the first power). Expressions can often be simplified by collecting like terms.

EXAMPLE 1

2y+2y is the same as $y \cdot (2+2)$, or 4y. Since the variables are identical, add the constants.

EXAMPLE 2

2x+y+3x+4y is the same as $[x \cdot (2+3)] + [y \cdot (1+4)]$ or 5x + 5y. In factored form, this is 5(x + y).

EXAMPLE 3

$x^2 + 3z + y$ cannot be factored or simplified further because there are no like terms and no common factors.

Multiplying Polynomials

In a polynomial expression, terms are added or subtracted. For example, 7x is not a polynomial, but 7x-3 is. When polynomials are multiplied together, the distributive property applies.

ALGEBRAIC PROPERTIES AND SIMPLE EQUATIONS

EXAMPLE

$(x+2)(y+3)$ does not equal xy + 5 or xy + 6. Instead every term in the first expression must multiply every term in the second. An easy way to do this is to take the first term of the first expression (positive x) and multiply every term in the second expression: $x(y+3) = xy+3x$. Now take the second term of the first expression (+2 or positive 2) and multiply it by every term in the second expression: $2(y+3) = 2y+6$. Finally, add the products: $xy+3x+2y+6$. Since there are no like terms, this is the final product of multiplying the two polynomials.

The same method applies to multiplying polynomials with more than two terms.

EXAMPLE

$(x+7)(x^2-3x+2)$ yields the following:

First, multiply x by all terms in the second expression:
$x \bullet (x^2-3x+2) = x^3 - 3x^2 + 2x$

Then, multiply 7 by all terms in the second expression:
$7 \bullet (x^2-3x+2) = 7x^2 - 21x + 14$

Add the products:
$x^3 - 3x^2 + 2x + 7x^2 - 21x + 14 = (x^3) + (-3x^2 + 7x^2) + (2x - 21x) + (14)$

Combine like terms to achieve the final result: $x^3 + 4x^2 - 19x + 14$.

COMMUTATIVE, ASSOCIATIVE, & DISTRIBUTIVE PROPERTIES QUESTIONS

a. Is a + b + c + d the same as c + a + d + b ?
b. Is $a \bullet 7 \bullet 3 \bullet b$ the same as $3 \bullet b \bullet 7 \bullet a$?
c. Is $(7+a)+(3+b)$ the same as $7+(a+3)+b$?
d. Is $(7a) \bullet (3b)$ the same as $3a \bullet (7b)$?
e. Is $7 \bullet (3b+2)$ the same as $(7+2) \bullet 3b$?
f. Is $9 \bullet (4x+2y)$ the same as $(36x)+(9 \bullet 2y)$?

CHAPTER 6

g. Use the distributive property to write: $x(3x+y+4z)$.

h. Factor the following: $5x + 10y - 120$.

i. Collect like terms: $25x+17y-12z-18x+3y$.

j. Collect like terms: $2xy-4x+18xy+16y-32x-30y+88xy-42x$.

k. Multiply these polynomials: $(x-y)(a+b)$.

l. Multiply these polynomials: $(12x-3y)(7x+4y-2z)$.

COMMUTATIVE, ASSOCIATIVE, & DISTRIBUTIVE PROPERTIES ANSWERS

a. Yes, because the commutative property of addition says that order does not matter.

b. Yes, because the commutative property of multiplication says that order does not matter.

c. Yes, because the associative property of addition says that grouping does not matter.

d. Yes, because the associative property of multiplication says that grouping does not matter.

e. No, because the associative property does not apply to mixed addition-multiplication expressions.

f. Yes, because the distributive property required that the nine multiply each of the two terms inside the parentheses. $9\bullet(4x+2y)=(9\bullet 4x)+(9\bullet 2y)=36x+18y$

g. $3x^2+xy+4xz$, because the x multiplied each expression inside the parentheses:
$x(3x+y+4z)=(x\bullet 3x)+(x\bullet y)+(x\bullet 4z)=3x^2+xy+4xz$

h. $5(x + 2y - 24)$, because $5x + 10y - 120$ is the same as
$(5\bullet x)+(5\bullet 2y)-(5\bullet 24)=5\bullet(x+2y-24)$.

i. $7x + 20y -12z$ because
$(25x-18x)+(17y+3y)+(-12z)=(7x)+(20y)+(-12z)=7x+20y-12z$

j. $108xy-78x-14y$, because $2xy-4x+18xy+16y-32x-30y+88xy-42x=$
$(2xy+18xy+88xy)+(-4x-32x-42x)+(16y-30y)=$
$(108xy)+(-78x)+(-14y)=108xy-78x-14y$

ALGEBRAIC PROPERTIES AND SIMPLE EQUATIONS

k. $xa + xb - ya - yb$, because x is multiplied by both terms in the second expression: $+x(a+b) = xa + xb$

Then $-y$ is multiplied by both terms in the second expression: $-y(a+b) = -ya - yb$

Finally, the products are added: $xa + xb - ya - yb$. Since there are no like terms to combine, the expression is completed.

l. $84x^2 - 12y^2 + 27xy - 24xz + 6yz$, because $12x$ is multiplied by all terms in the second expression: $+12x(7x + 4y - 2z) = 84x^2 + 48xy - 24xz$

Then $-3y$ is multiplied by all terms in the second expression: $-3y(7x + 4y - 2z) = -21xy - 12y^2 + 6yz$

Finally, the products are added: $(84x^2) + (-12y^2) + (48xy - 21xy) + (-24xz) + (6yz)$

And like terms are combined: $84x^2 - 12y^2 + 27xy - 24xz + 6yz$

Order of Operations

The order of operations was introduced in the chapter on calculators. Here, the hierarchy will be put to use.

EXAMPLE 1

To simplify $16 - 4 \bullet 3^2 + \frac{12}{2}$, we use the "PEMDAS" sequence (Parentheses, Exponents, Multiplication, Division, Addition, and Subtraction). Since there are no Parentheses, skip the "P." Since there is an Exponent (3^2), work that next to get nine. The expression now looks like this: $16 - 4 \bullet 9 + \frac{12}{2}$. Next, perform Multiplication and Division ($4 \bullet 9$ and $\frac{12}{2}$) to obtain a 36 and a six. The expression now looks like this: $16 - 36 + 6$. Finally, perform Addition and Subtraction to get the answer: -14.

CHAPTER 6

EXAMPLE 2

$5 \bullet [(18-6)+8]$ simplifies to 100, because

First, the inner parentheses (18-6) must be simplified to 12: ($5 \bullet [(12)+8]$)

Second, because the middle parentheses (12+8) must be simplified to 20: $5 \bullet (20)$. And finally, simplify to 100, because the multiplication must be performed last.

EXAMPLE 3

$12 - \left[4^2 + \left(\frac{10}{2} - 8 \right) \bullet \frac{4}{8} \right]$ simplifies to -2.5: First, simplify the inner parentheses

$\left(\frac{10}{2} - 8 \right) = (5-8) = -3$, so now the expression is $12 - \left[4^2 + (-3) \bullet \frac{4}{8} \right]$.

Second, simplify the term with the exponent (4^2) to 16. Now the expression is:

$12 - \left[16 + (-3) \bullet \frac{4}{8} \right]$

Third, perform the multiplication and division: $\left(-3 \bullet \frac{4}{8} \right) = (-3 \bullet 0.5) = -1.5$

The expression now reads: $12 - (16 - 1.5)$

Fourth, perform the subtraction inside the parentheses: (16-1.5) = 14.5. The expression now reads 12– (14.5)

Fifth, perform the final subtraction 12–14.5, giving an answer of -2.5.

Rules of Equations

Multiplying by one does not change the value.

You can multiply *either side* or *both sides* of an equation by one without changing the value.

ALGEBRAIC PROPERTIES AND SIMPLE EQUATIONS

EXAMPLE 1

In the equation $3 = \frac{6}{2}$, you can multiply either side by any value equal to one without invalidating the equation. For example, $3 = \left(\frac{6}{2}\right) \cdot \left(\frac{10}{10}\right)$ is the same as $3 = \frac{60}{20}$. This has not changed the validity of the equation.

EXAMPLE 2

The equation $x = \frac{-3}{-8}$ can be changed by multiplying the fraction side of the equation (only) by a one in the form of $\frac{-1}{-1}$. The equation now reads $x = \left(\frac{-1}{-1}\right)\left(\frac{-3}{-8}\right)$, or $x = \frac{3}{8}$.

You may do any of the following to both sides of an equation without changing its validity:

> **Add the same term**
>
> **Subtract the same term**
>
> **Multiply by the same term**
>
> **Divide by the same term as long as it's not zero**

EXAMPLE 1

In the equation $x = 3$, $x + 4 = 3 + 4$ is still a valid equation. Adding four to both sides does not change the equality, or the values (x still must equal three for the equation to balance).

EXAMPLE 2

$x - 4 = 3 - 4$ is also a valid equation. Subtracting four from both sides did not change the equality, or the values (x still must equal three for the equation to balance).

EXAMPLE 3

$x \cdot 4 = 3 \cdot 4$ is also a valid equation. Multiplying both sides by four did not change the equality, nor the values (x still must equal three for the equation to balance).

EXAMPLE 4

$\frac{x}{4} = \frac{3}{4}$ is also a valid equation. Dividing both sides by four did not change the equality or the values (x still must equal three for the equation to balance).

EXAMPLE 5

Note that for the equation $x = 3$, $x = 3 + 4$ and $x + 4 = 3$ are not valid because the addition was performed on only one side of the equation.

$x = 3 - 4$ and $x - 4 = 3$ are not valid because the subtraction was performed on only one side of the equation.

$x = 3 \cdot 4$ and $4x = 3$ are not valid because the multiplication was performed on only one side of the equation.

$x = \frac{3}{4}$ and $\frac{x}{4} = 3$ are not valid because the division was performed on only one side of the equation. Additionally: $\frac{x}{0} = \frac{3}{0}$ is not valid because division by zero is not allowed.

Solving Equations with a Variable on One Side

Rules of equations are frequently used to isolate variables. In the following examples, rules of equations are used to isolate the variable on one side of the equation and all of the numeric terms on the other side.

ALGEBRAIC PROPERTIES AND SIMPLE EQUATIONS

EXAMPLE

To isolate the variable in the equation $x + 3 = 5$, subtract three from both sides of the equation: $x + 3 - 3 = 5 - 3$. Thus, $x + 0 = 2$, or $x = 2$.

In addition to using addition, subtraction, multiplication, and division, exponents and radicals may also be used to isolate a variable.

EXAMPLE

To isolate the variable in the equation $\sqrt{x} = 4$, both sides must be squared: $(\sqrt{x})^2 = (4)^2$. Since the square of a square root is the radicand itself, the equation simplifies to $x = 4^2$ or $x = 16$.

SINGLE VARIABLE QUESTIONS

a. Solve for x: $x + 5 = 20$

b. Solve for x: $x - 8 = 4$

c. Solve for x: $x \bullet 5 = 30$

d. Solve for x: $\dfrac{9}{x} = 3$

e. Solve for x: $3x^3 = 81$

f. Solve for x: $4 - 5x^2 = -121$

SINGLE VARIABLE ANSWERS

a. $x = 15$ because $x + 5 - 5 = 20 - 5$ which simplifies to $x + 0 = 15$ or $x = 15$.

b. $x = 12$, because $x - 8 + 8 = 4 + 8$ which simplifies to $x - 0 = 12$ or $x = 12$.

c. $x = 6$, because $\dfrac{x \bullet 5}{5} = \dfrac{30}{5}$ which simplifies to $x \bullet \dfrac{5}{5} = 6$ or $x \bullet 1 = 6$ or $x = 6$.

d. $x = 3$ Multiply both sides by x: $\dfrac{9}{x} \bullet x = 3 \bullet x$. This simplifies to $9 = 3x$. Then divide both sides by 3 to isolate the x: $\dfrac{9}{3} = \dfrac{3x}{3}$. This leaves $x = \dfrac{9}{3}$ or 3.

e. $x = 3$ First, divide both sides by three: $\frac{3x^3}{3} = \frac{81}{3}$. When this is simplified, $x^3 = 27$ is the result. To obtain a value for x, take the cube-root of both sides: $\sqrt[3]{x^3} = \sqrt[3]{27}$. When this expression is simplified, we are left with $x = 3$. Since the order of the radical is odd, and the radicand is positive, only a positive three is valid.

f. $x = \pm 5$ First, subtract four from both sides to isolate $-5x^2$: $-5x^2 + 4 - 4 = -121 - 4$. This simplifies to $-5x^2 = -125$. Next, divide both sides of the equation by -5 to isolate the x^2: $\frac{-5x^2}{-5} = \frac{-125}{-5}$. When this is simplified, $x^2 = 25$ remains. Note that dividing a negative by a negative gives a positive result. To isolate the variable now, take the square root of both sides: $\sqrt{x^2} = \sqrt{25}$, so $x = \pm 5$, because even-order radicals have two valid roots.

Solving Equations with Variables on Both Sides

Often, variables appear on both sides of an equation. In order to isolate the variable, all instances of the variable must be moved to one side using the rules of equations stated previously.

EXAMPLE 1

$4x + 7 = 3x - 5$. In order to isolate the variable, all constants must be moved to one side of the equation and all variables to the other. Which side of the equal sign the variables are moved to does not matter, though by mathematical convention, they are isolated to the left. In this example, you will move the variables to the left side of the equal sign.

First, move the variable from the right side of the equation to the left side by subtraction: $4x + 7 - 3x = 3x - 5 - 3x$. Combine terms: $(4x - 3x) + 7 = (3x - 3x) - 5$. This simplifies to $x + 7 = -5$. To further isolate the variable, subtract seven from both sides: $x + 7 - 7 = -5 - 7$. This simplifies to $x = -12$.

EXAMPLE 2

Isolate x: $3x + 7 - (6 \bullet 2x) + 4 = 17x(3 + 2^2)$.

First, simplify the expressions in parentheses and the exponents:
$3x + 7 - (12x) + 4 = 17x(3 + 4)$.

Next, group terms: $(3x - 12x) + (7 + 4) = 17x(7)$. This simplifies to $(-9x) + 11 = 119x$.

Next, isolate x on one side by adding $9x$ to both sides: $-9x + 11 + 9x = 119x + 9x$. This simplifies to $11 = 128x$.

Finally, to isolate the variable, divide both sides by 128: $\frac{11}{128} = \frac{128x}{128}$.

This leaves $x = \frac{11}{128}$, or $x = 0.086$, approximately.

EXAMPLE 3

Isolate x: $3x^2 - 48 = x^2 + 12$.

Subtract x^2 from both sides to put all of the x^2 terms on the left side: $3x^2 - 48 - x^2 = x^2 + 12 - x^2$.

Combine terms: $(3x^2 - x^2) - 48 = (x^2 - x^2) + 12$, or $2x^2 - 48 = 12$.

Now isolate the variable by adding 48 to both sides: $2x^2 - 48 + 48 = 12 + 48$. Combine terms to get: $2x^2 = 60$.

Next, divide both sides by two to isolate the x^2 term: $\frac{2x^2}{2} = \frac{60}{2}$, or $x^2 = 30$.

Finally, to isolate the variable, take the square root of both sides: $\sqrt{x^2} = \sqrt{30}$. The numerical equivalent is $x = \pm 5.477$, approximately. Note that since the order was even, there are two roots—a positive and a negative.

Solving Equations with Multiple Variables

Some equations will have more than one variable. In these cases, select one variable, and solve for it in terms of the second variable.

EXAMPLE

$4x+3=22y$ can be solved for either x or y, but not both. Either answer will contain the second variable. Solving for x (subtracting three from both sides and dividing both sides by four), the result is $x = \frac{22y-3}{4}$. Solving for y (dividing both sides by 22), the result is $y = \frac{4x+3}{22}$. When setting up equations, single-variable equations are almost always preferable to multiple-variable expressions.

Solving Simultaneous Equations

Sometimes, multiple equations are found that must be solved simultaneously (for example, two equations and two unknowns that must be solved for). These often contain multiple variables. The easiest way to solve simultaneous equations is to use the calculator if your calculator offers this option. If your calculator does not have a simultaneous equations mode, use one of the following methods:

Method 1: Solve the first equation for one of the variables.

Then plug the answer (in terms of the second variable) into the second equation. This produces an equation with only one variable. Solve for that variable, plug that value back into the first equation, and solve for the original variable.

EXAMPLE

Given the equations $x+2y=17$ and $x-y=2$, solve the first equation ($x+2y=17$) for x: $x+2y-2y=17-2y$, which simplifies to $x=17-2y$.
Now plug this value for x into the second equation where the x-variable appears: $(17-2y)-y=2$.
Combine terms: $17-3y=2$.
Then isolate the y term: $17-3y-17=2-17$ which becomes $-3y=-15$.
Divide both sides by -3 to get $y=5$.
Once the y value has a numeric equivalent (five), plug that number back into the original equation ($x+2y=17$) to solve for x: $x+2\bullet(5)=17$, which simplifies to $x+10=17$.
Subtract 10 from both sides of the equation to find that $x = 7$. The answer set for

this pair of equations is $x = 7$ and $y = 5$.
To check these numbers, plug them into both of the original equations and make sure that both equations balance.

Method 2: Add the equations and solve the equation formed by the sum of the two.

To use this method, multiply both sides of either equation by a number chosen so that one of the variables cancels out.

EXAMPLE

Using the same equations as in the first method ($x + 2y = 17$ and $x - y = 2$), we note that the *y*-value is subtracted in the second equation. To make the *y*'s cancel out, we can multiply the second equation (both sides) by two so that the $+2y$ from the first equation will sum to zero with the $-2y$ from the second. This makes the second equation: $2 \bullet (x - y) = 2 \bullet 2$, which simplifies to $2x - 2y = 4$.
Then add both equations as follows:

$$x + 2y = 17$$
(+) $$2x - 2y = 4$$

$$3x = 21$$

Divide both sides by three to isolate the variable, and the result is $x = 7$.

Once the *x*-variable is defined, substitute that value back into either original equation to find that $y = 5$.

Method 3: Since the certification examinations are multiple-choice tests, take each pair of answers and plug them into the given equations. One and only one answer will satisfy both equations.

This option is deliberately not available in this book to ensure that you are exposed to proper problem-solving methods.

CHAPTER 6

MULTIPLE VARIABLES AND MULTIPLE EQUATIONS QUESTIONS

a. Solve for x: $3x + 9 = 7(6 + 2x)$

b. Solve for x: $3(x^2 - 2) - 14 = \dfrac{4x^2}{6}$

c. Solve for x: $7x - 12 = 4(y + 3)$

d. Solve for x: $-4(y - 3) = \sqrt{y + x}$

e. Solve these equations: $x + y = 56$ and $497 = 7x + 12y$

MULTIPLE VARIABLES AND MULTIPLE EQUATIONS ANSWERS

a. -3, because

Multiply and simplify:	$3x + 9 = 42 + 14x$
Collect terms:	$3x + 9 + (-14x) + (-9) = 42 + 14x + (-14x) + (-9)$
The result is:	$3x - 14x = 42 - 9$ or $-11x = 33$.
Divide both sides by -11:	$x = -\dfrac{33}{11}$ or x = -3.

b. $x = \pm 2.928$, approximately, because

Multiply terms in parentheses first:	$3x^2 - 6 - 14 = \dfrac{4x^2}{6}$
Combine terms:	$3x^2 - 20 = \dfrac{4x^2}{6}$
Multiply both sides by 6:	$6(3x^2 - 20) = 4x^2$
The result is:	$18x^2 - 120 = 4x^2$
Subtract $4x^2$ from both sides to move variables to the left of the equals sign:	$18x^2 - 4x^2 - 120 = 0$
Combine terms and add 120 to both sides:	$14x^2 = 120$
Divide both sides by 14:	$x^2 = \dfrac{120}{14} = 8.5714285$

ALGEBRAIC PROPERTIES AND SIMPLE EQUATIONS

	Take the square root of both sides to bring x to the first power:	$\sqrt{x^2} = \sqrt{8.5714285}$
	The result is:	$x = \pm 2.928$, approximately
c.	$x = \dfrac{4y+24}{7}$, because	
	Multiply terms in parentheses:	$7x - 12 = 4y + 12$
	Add twelve to both sides of the equation to isolate x on the left:	$7x - 12 + 12 = 4y + 12 + 12$
	Combine terms and divide both sides by seven to isolate the x:	$x = \dfrac{4y+24}{7}$
d.	$x = 16y^2 - 97y + 144$, because	
	Multiply terms in parentheses:	$-4y + 12 = \sqrt{y+x}$
	Square both sides to eliminate the radical on the right side of the equation.	$(-4y+12)^2 = \left(\sqrt{y+x}\right)^2$
	The result is:	$16y^2 - 48y - 48y + 144 = y + x$
	Combine terms and subtract y from both sides to isolate y values on the left side of the equation	$16y^2 - 97y + 144 = x$.
e.	$x = 35$ and $y = 21$, because	
	Solve the first equation for x:	$x = 56 - y$
	Plug that value in for x in the second equation:	$497 = 7(56 - y) + 12y$
	Perform multiplication:	$497 = 392 - 7y + 12y$
	Combine terms and subtract 392 from both sides	$105 = 5y$
	Divide both sides by five to isolate y:	$21 = y$

Plug the y value back into the first
equation and solve for x: $\quad x+21=56.$

Subtract twenty-one from both
sides to isolate the x: $\quad x=35$

Proportions

In the chapter on fractions, a proportion was an equation where the cross-products of the fractions on either side of the equal sign were identical.

EXAMPLE

$\frac{3}{7}=\frac{9}{21}$ is a proportion because $3 \bullet 21 = 9 \bullet 7$. (Both sides equal 63).

Direct proportions are represented as division in the form of fractions.

The proportion concept can be used to verify or disprove the equivalence of values.

EXAMPLE

You can buy two filters for $37 or ten for $185. Are you paying the same price per filter either way?

$$\begin{array}{l} Parts \rightarrow \\ Dollars \rightarrow \end{array} \frac{2}{37} = \frac{10}{185} \begin{array}{l} \leftarrow Parts \\ \leftarrow Dollars \end{array}$$

To answer this question, cross multiply: $2 \bullet 185 = 370$ and $37 \bullet 10 = 370$. Since the products of cross multiplication are the same (370) the prices are identical, and the equation is a proportion.

The concept becomes even more powerful when one of the values is unknown.

EXAMPLE

If you can buy 21 parts for $252, what would 37 parts cost?

Use the same style to set up the equation: $\begin{array}{l} Parts \rightarrow \\ Dollars \rightarrow \end{array} \frac{21}{252} = \frac{37}{x} \begin{array}{l} \leftarrow Parts \\ \leftarrow Dollars \end{array}$

To determine the value of x,	
cross multiply:	$21x = 37 \bullet 252$ or $21x = 9324$.
To isolate the x value,	
divide both sides by 21:	$\dfrac{21x}{21} = \dfrac{9324}{21}$
Complete the division:	$x = \$444$.

Reality Check! Whenever you use a proportion, look at the original fraction and the answer to see if your answer makes sense. If the numerator on the right side of the equation got larger, then the denominator must also be larger for the fractions to be proportional. Since the numerator in this example increased—from 21 to 37, the denominator must also increase (a *proportional* amount) from the original 252. Since the answer (444) is, indeed, larger than 252, the answer seems credible.

Since the denominator increases as the numerator increases, this is called a *direct* proportion. Questions often use the terms "is directly proportional to," "varies directly as to…," and "changes as to" to indicate a direct proportion relationship. Use a direct proportion to solve a problem that contains any of these descriptions.

Inverse proportions are represented as multiplications.

Questions frequently use the terms "inversely proportional," and "varies inversely as to" to indicate the opposite of a direct proportion—an *inverse* proportion. Since direct proportions are set up as divisions, inverse proportions are set up as multiplications.

EXAMPLE

The weight needed to balance a see-saw varies inversely with the distance from the center in feet. If a 50-pound child is sitting 4 feet from the center of the see-saw, how far from the center must a 40-pound child sit for the see-saw to balance?

The *Candidates' Handbook* gives this equation as: $F_1 D_1 = F_2 D_2$, where the "F" variable represents force and the "D" variable represents distance. For the problem at hand, the "D_2" variable is the one sought:
$50(pounds) \bullet 4(feet) = 40(pounds) \bullet D_2(feet)$ works out to $200\, ft.lbs. = 40 lbs \bullet D_2$.

To solve for the unknown, divide both sides by 40 pounds:

$$\frac{200\ ft.lbs.}{40(pounds)} = \frac{40(pounds) \cdot D_2(feet)}{40(pounds)}.$$

The "pounds" units cancel, and 200 divided by 40 produces $5(feet) = D_2(feet)$.

Reality check! Whenever you use an inverse proportion, look at the original number and the answer. Since the pounds *decreased* between the left side of the equation and the right, (from 50 to 40), the distance *must increase* (a *proportional* amount) for the equation to remain balanced. Since the new distance (five feet) is greater than the original distance (four feet) the answer seems credible. Since one value decreases as the other increases, this is an *inverse* proportion.

You will not always be provided with a formula. You may have to determine for yourself whether to divide or multiply based on the language of the problem.

PROPORTIONS QUESTIONS

a. Your company earns $728 for 50 units of product produced. How much will be earned for 78 units?

b. There is an inverse relationship between a site's score on safety inspections and the site's number of recordable injuries. If a site with a score of 28 on its safety inspection has 7 recordable injuries, how many injuries would be expected for a site that scored 49 on its safety inspection?

c. If one company offers four minutes of satellite time for $7,400, and another company offers twenty minutes for $37,000, are the minutes priced equally?

d. The length of a wire is directly proportional to its resistance. If a 20-foot length of wire has a resistance of 4 ohms, how much resistance does a 22-foot length have?

e. The strength of an acid and the gallons to be used vary inversely. If you must use two gallons of 28% acid to achieve the proper strength, how many gallons of 20% acid must you use?

PROPORTIONS ANSWERS

a. $\frac{Dollars \rightarrow}{Units \rightarrow} \frac{728}{50} = \frac{x}{78} \frac{\leftarrow Dollars}{\leftarrow Units}$ is the setup because the cost will be proportional.

Cross-multiply: $728 \bullet 78 = 56{,}784$ and $50 \bullet x = 50x$.

Set the products equal: $56{,}784 = 50x$.

Divide both sides by 50 to isolate x: $\dfrac{56{,}784}{50} = \dfrac{50x}{50}$, or $x = \$1{,}135.68$

b. Since this is an inverse relationship, set up the problem as a multiplication equation with the original score times original injuries equal to the new score times new injuries (the "x" to solve for): $28(score) \bullet 7(injuries) = 49(score) \bullet x(injuries)$.

Multiply: $196 = 49x$

To isolate the x, divide both sides by 49: $\dfrac{196}{49} = \dfrac{49x}{49}$

This leaves: $4 = x$, the number of injuries expected.

c. Yes, because $\dfrac{Minutes \rightarrow}{Dollars \rightarrow} \dfrac{4}{7{,}400} = \dfrac{20}{37{,}000} \dfrac{\leftarrow Minutes}{\leftarrow Dollars}$ can be cross multiplied as follows: $4 \bullet 37{,}000 = 20 \bullet 7{,}400$. Both cross products are equal to 148,000. Since the equation balances, this is a proportion, and the minutes are equally priced.

d. 4.4 ohms. Because you are told that this is a direct proportion, set up the equation with a pair of fractions: $\dfrac{Length \rightarrow}{Ohms \rightarrow} \dfrac{20}{4} = \dfrac{22}{x} \dfrac{\leftarrow Length}{\leftarrow Ohms}$.

Cross-multiply: $20x = 4 \bullet 22$

Simplify: $20x = 88$

Divide both sides by 20: $x = 4.4$

Since the resistance given was in ohms, the answer is 4.4 ohms.

e. 2.8 gallons. Because you are told that this is an inverse proportion, the gallons and strength must be multiplied together.

$2 gallons \bullet 28\% strength = ? gallons \bullet 20\% strength$. Notice that since the strength is decreasing, the number of gallons must increase.

The equation becomes: $2 \bullet 28 = 20x$ or $56 = 20x$

Divide both sides by 20: $2.8 = x$.

Reality check! Since you know that you need more gallons of 20% than 28%, the number of gallons in the answer must be larger than the original 2 gallons. Since the answer, 2.8, is indeed larger, it seems to be a credible answer. If the answer had been less than 2 gallons, you would know that something is wrong.

ALGEBRAIC PROPERTIES AND SIMPLE EQUATIONS CHALLENGE EXAM

a. Is the following equation valid? $-x^2 - 7x + 4y - 3z = 4y - x^2 - 7x - 3z$

b. Is the following equation valid? $x^2 \bullet (-3) \bullet (-y) \bullet 2 = 6x^2 y$

c. Is the following equation valid? $[(7-3)+(4-6)]+2 = 2+7-3+4-6$

d. Is the following equation valid? $3(xy) \bullet 7(z) = 21xyz$

e. Is the following equation valid? $x(4+y) = 4x(y)$

f. Factor the following expression: $(4x + 8xy - 16xz^2)$

g. Simplify this expression: $3x^2 + 4x(x+y) - 3xy + y^2 - 2x^2$

h. Simplify this expression: $(7-x)(x+y)$

i. Simplify this expression: $(x+y+z)(2x-3y+4z)$

j. Find the numeric value of this expression: $8 + \left[2^3 - \left(\dfrac{7}{14} - 2 \right) \bullet \dfrac{18}{6} \right]^2$

k. Solve for x: $\dfrac{x}{7} + 3 = 17$

l. Solve for x: $5x - 4 + (2 \bullet 3x) + 5 = x(4-2)^2$

m. Solve these equations for x and y values: $2x - 3y = 13$ and $-3x + 5y = -21$

n. The resistance of a wire is directly proportional to its length. If a 20-foot wire has a resistance of 4 ohms, what resistance will a 28-foot section have?

o. The strength of a caustic solution and the number of gallons needed to produce it are inversely proportional. If 4 gallons of 25% caustic solution are required per batch, how many gallons of 30% solution will be needed for a batch of the same strength?

p. The resistance of wire is directly proportional to its length and inversely proportional to the square of its diameter. A 20-foot length of 2mm diameter wire has a resistance of 6 ohms. What will be the resistance of a 35-foot length of 3mm diameter wire?

ALGEBRAIC PROPERTIES AND SIMPLE EQUATIONS CHALLENGE EXAM ANSWERS

a. Yes, because the commutative property shows that order does not matter for addition or subtraction.

b. Yes, because the commutative property shows that order does not matter for multiplication. Also, the two negative values become positive when multiplied.

c. Yes, because the associative property shows that grouping does not matter for addition or subtraction.

d. Yes, because the associative property shows that grouping does not matter for multiplication.

e. No, because the distributive property says that a term that is multiplied by a polynomial must be multiplied by each term of the polynomial. The correct answer is $x(4+y)=4x+4y$. The equation as shown yields a wrong answer of $4xy$.

f. $4x(1+2y-4z^2)$, because the term $4x$ is common to all of the terms of the polynomial inside the parentheses.

g. $5x^2 + xy + y^2$, because

Perform multiplication first, yielding: $\quad 3x^2 + 4x^2 + 4xy - 3xy + y^2 - 2x^2$

Combine like terms: $\quad (3x^2 + 4x^2 - 2x^2) + (4xy - 3xy) + y^2 = 5x^2 + xy + y^2$

h. $7x + 7y - x^2 - xy$, because each term of the first polynomial must be multiplied by each term of the second, and then the products must be added.

Multiply by seven: $\quad 7(x+y) = 7x+7y$

Multiply by $-x$: $\quad -x(x+y) = -x^2 - xy$

Add the two products: $\quad 7x + 7y - x^2 - xy$

Since this cannot be simplified any further (no common terms for factoring and no like terms to add or subtract), this is the final answer.

CHAPTER 6

i. $2x^2 - 3y^2 + 4z^2 - xy + 6xz + yz$, because each term of the first polynomial must be multiplied by every term of the second, and then the products must be added.

Multiply the x from
the first polynomial: $\quad x(2x - 3y + 4z) = 2x^2 - 3xy + 4xz$

Multiply the y from
the first polynomial: $\quad y(2x - 3y + 4z) = 2xy - 3y^2 + 4yz$

Multiply the z from
the first polynomial: $\quad z(2x - 3y + 4z) = 2xz - 3yz + 4z^2$

Add these three products: $2x^2 - 3xy + 4xz + 2xy - 3y^2 + 4yz + 2xz - 3yz + 4z^2$

Combine the terms:
$2x^2 + (-3xy + 2xy) + (4xz + 2xz) + (4yz - 3yz) - 3y^2 + 4z^2 = 2x^2 - 3y^2 + 4z^2 - xy + 6xz +$

j. 164.25, because hierarchy of operations must be used to simplify the expression.

Resolve parentheses
from the inside out:
$$8 + \left[2^3 - (0.5 - 2) \bullet \frac{18}{6}\right]^2 = 8 + \left[2^3 - (-1.5) \bullet \frac{18}{6}\right]^2$$

Now that the inner parentheses have a numeric equivalent, the exponents within the outer parentheses must be satisfied:
$$8 + \left[8 - \left(-1.5 \bullet \frac{18}{6}\right)\right]^2.$$

Now that exponents are satisfied, multiply and divide inside the parentheses:
$$8 + [8 - (-1.5 \bullet 3)]^2 = 8 + (8 + 4.5)^2.$$

Now that multiplication and division inside the parentheses are complete, add and subtract inside the parentheses:
$$8 + (12.5)^2$$

Since what's inside the parentheses now has a numeric answer, the process begins again with the parentheses' exponent being satisfied: $\quad 8+(156.25)=164.25$

k. x = 98, because

To isolate x, subtract three from both sides of the equation: $\quad \dfrac{x}{7}+3-3=17-3$

This yields: $\quad \dfrac{x}{7}=14$

Multiply both sides by 7: $\quad 7\left(\dfrac{x}{7}\right)=7\bullet 14$, or x = 98

l. Approximately –0.143, because:

Simplify what's inside the parentheses: $\quad 5x-4+(6x)+5=x(2)^2$

Find values of terms with exponents and combine like terms: $\quad (5x+6x)-4+5=4x$

Simplify: $\quad 11x+1=4x$

Subtract 4x from both sides: $\quad 11x+1-4x=4x-4x$

Subtract 1 from both sides: $\quad 7x+1-1=-1$

Divide both sides by 7: $\quad \dfrac{7x}{7}=\dfrac{-1}{7}$ or x = –0.143 (approximately)

m. x = 2 and y = -3, because, first, solve the first equation ($2x-3y=13$) for x:

Add *3y* to both sides: $\quad 2x=13+3y$

Divide both sides by 2: $\quad x=\dfrac{13+3y}{2}$

Plug this expression into the second equation in place of the *x*: $\quad -3\left(\dfrac{13+3y}{2}\right)+5y=-21$

CHAPTER 6

Simplify by performing the multiplication
on the parentheses first:
$$\frac{-39-9y}{2}+5y=-21$$

Perform the addition on the left,
but find a common denominator first:
$$\frac{-39-9y}{2}+\left(\frac{2}{2}\cdot\frac{5y}{1}\right)=-21$$

This produces:
$$\frac{-39-9y+10y}{2}=-21$$

Multiply both sides by two and combine terms: $-39+y=-42$

Add 39 to both sides: $y=-3$

Plug this value for y back into one of the
original equations and solve for x: $2x-3\bullet(-3)=13$

Perform the multiplication: $2x+9=13$

Subtract nine from both sides: $2x=13-9=4$

Divide both sides by two: $x=2$

Test the solution set ($x = 2$, $y = -3$) by using these values to verify both of the original equations.

n. $x = 5.6$ ohms. Because this is a direct proportion, the problem is set up as an equation using fractions: $\dfrac{\text{Length} \rightarrow}{\text{Resistance} \rightarrow}\ \dfrac{20}{4}=\dfrac{28}{x}\ \dfrac{\leftarrow \text{Length}}{\leftarrow \text{Resistance}}$

Cross-multiply: $20x = 112$

Divide both sides by 20: $x = 5.6$.

o. 3.33 gallons. Because this is an inverse proportion, the volume (gallons) must decrease as the strength increases. The original solution was 25% caustic, and the new solution is stronger (30%), so fewer gallons must be used.

Set up the inverse proportion
as a multiplication problem: $(4)\bullet(25\%)=(x)\bullet(30\%)$

Perform the multiplication: $100 = 30x$

Divide both sides by 30 to isolate x: $x = 3.33$ or three and a third gallons

ALGEBRAIC PROPERTIES AND SIMPLE EQUATIONS

Since this is less than the 4 gallons of weaker caustic that you started with, it seems a plausible answer.

p. 4.67 ohms. Because this problem contains both direct and inverse proportions, a combination of fractions and multiplication is used to solve it.

Deal with the direct proportion first: $\dfrac{Resistance \rightarrow}{Length \rightarrow} \quad \dfrac{6}{20} = \dfrac{x}{35} \quad \dfrac{\leftarrow Resistance}{\leftarrow Length}$

Note that either resistance or length can be used as the denominator, so long as the two sides are consistent.

Include the inverse proportion to the equation: $\quad 2^2 \cdot \dfrac{6}{20} = 3^2 \cdot \dfrac{x}{35}$

Since the resistance was inversely proportional to the *square* of the diameter (in mm), both diameters are squared. Solve for the variable: $\quad 4 \cdot \dfrac{6}{20} = 9 \cdot \dfrac{x}{35}$,

$$\text{or } \dfrac{4}{1} \cdot \dfrac{6}{20} = \dfrac{9}{1} \cdot \dfrac{x}{35},$$

$$\text{or } \dfrac{24}{20} = \dfrac{9x}{35}$$

Cross-multiply: $\quad 24 \cdot 35 = 20 \cdot 9x$, or $840 = 180x$

Divide both sides by 180: $\quad 4.67 = x$, approximately

CHAPTER SEVEN

Applied Algebra

Assumptions: You have sufficient knowledge to take and pass the challenge exam of all previous chapters. You have a scientific calculator and basic knowledge of how to use it.

Application: Required for ASP, CSP, OHST, and CHST examinations.

Discussion: Understanding algebra is essential to practical mathematics. The ability to solve equations by algebraic manipulation is a skill that will be required in both Safety Certification examinations and in real life. If you feel that your comprehension of algebra is good, go to the challenge exam at the end of this chapter. If you can work all of the problems correctly, skip this chapter.

Sets, Subsets, and Venn Diagrams

A set is a collection of things.

{2,4,6,8} represents a set consisting of the numbers 2, 4, 6, and 8. Things in a set are usually separated by commas and enclosed in braces to identify them as a set.

Each item in a set is called an "element."

The elements of the set shown above would be the actual numbers two, four, six, and eight.

Capital letters are often used to name sets.

A = {1,2,3} might be described in a problem as "set A."

A subset exists when all elements of the subset are also elements of the parent set.

EXAMPLE 1

Consider the following sets: A = {2,4,6,8} and B = {2,6} Set B is a subset of set A because all elements of B are also elements of A.

EXAMPLE 2

Given the following sets: C = {2,4,6,8} and D = {2,4,5,6}, D is not a subset of C because one element of set D (the five) is not an element of set C.

Set Symbols

Although candidates are unlikely to encounter set symbols on certification examinations, it is possible. The symbols commonly used with sets are given here.

Symbol	Description
$\subset$	"Is a subset of" so that for sets A = {1,3,5,7} and B = {3,5}, B $\subset$ A.
$\subseteq$	"Is a subset of or equal to" so that for sets A = {1,3,5} and C = {1,3,5}, A $\subseteq$ C and C $\subseteq$ A
$\not\subset$	"Is not a subset of" so that for sets A = {1,3,5} and D = {5,7,9}, A $\not\subset$ D and D $\not\subset$ A
$\cap$	"The intersection of" so that for sets A = {1,3,5} and D = {5,7,9}, A $\cap$ D = {5}
$\cup$	"The union of" so that for sets A = {1,3,5} and D = {5,7,9}, A $\cup$ D = {1,3,5,7,9}
$\in$	"Is an element of" so that for set A = {1,3,5}, 1 $\in$ A
$\notin$	"Is not an element of" so that for set A = {1,3,5}, 2 $\notin$ A

Table 11: Set Symbols

APPLIED ALGEBRA

Venn diagrams are a way of diagramming the relationship between sets.

EXAMPLE

Given the following sets: A = {2,4,6,8}, B = {2,4} and C = {2,4,5,7}

Venn diagrams showing the relationship between set A and set B would be as follows:

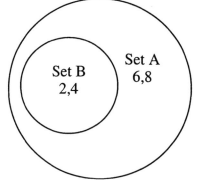

All elements inside the circle of set A (2,4,6, and 8) are elements of set A. All elements inside the circle of set B (2, and 4) are elements of both sets A and B.

The Venn diagram below shows the relationships between sets A {2,4,6,8}, B {2,4}, and C {2,4,5,7} as shown with a Venn diagram, becomes:

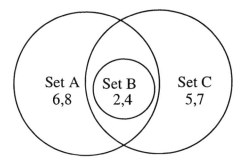

All elements inside the circle of set A (2,4,6, and 8) are elements of set A. All elements inside the circle of set C (2,4,5, and 7) are elements of set C. The overlapping area (called the intersection) contains elements that are common to

105

both sets A and C (elements 2 and 4). Set B contains only elements that happen to be in the intersection of sets A and C.

Venn diagrams are often used to identify mutually exclusive events. The study of probability requires these concepts.

Mixtures

Mixture problems are common in real life as well as on safety certification examinations. Mixture problems are most commonly simple inverse proportion problems (as the amount gets larger, the concentration must become less to maintain the same overall amount of ingredient, and vice versa). Some examples:

EXAMPLE 1

Starting with 10 gallons of 50% solution, how many gallons of 25% solution must be added to make the overall concentration 40%?

Because the unknown you seek is the number of gallons of 25% solution, that will become the "x." Diagraming the problem mathematically:

$$(10\,gal. \bullet 50\%) + (x \bullet 25\%) = (x + 10\,gal.) \bullet 40\%$$

Note that the volume of 40% mixture had to be the sum of the original 10 gallons and the x-gallons of additive.

Convert percent to decimal form:	$(10 \bullet 0.5) + 0.25x = 0.4x + (10 \bullet 0.4)$
Perform multiplication:	$5 + 0.25x = 0.4x + 4$
Subtract $0.25x$ from both sides:	$5 = 0.15x + 4$
Subtract 4 from both sides:	$1 = 0.15x$
And divide both sides by 0.15:	$6.667 = x$ gallons (approximately)

EXAMPLE 2

Twelve pints of an unknown concentration solution are added to fifteen pints of 50% concentration. The resulting mixture tests as 42% concentration. What was the concentration of the unknown solution?

Because the variable you seek is the concentration of the unknown, that will become the "x." Diagraming the problem:

$$(12 pts. \bullet x) + (15 pts. \bullet 50\%) = 27 pts. \bullet 42\%$$

Convert percent to decimal:	$12x + (15 \bullet 0.5) = 27 \bullet 0.42$
Perform multiplication:	$12x + 7.5 = 11.34$
Subtract 7.5 from both sides:	$12x = 3.84$
And divide both sides by 12:	$x = 0.32$ (or 32%)

Both of the mixture examples above involved only one variable and one equation. The following is an example using two variables and two equations:

EXAMPLE

A reactor requires 250 gallons of 50% initiator to begin its reaction. The technicians have only 25% and 65% solutions on hand. How much of each must be mixed to provide the 250 gallons of 50% initiator needed?

For this case, allow "x" to represent the gallons of 25% solution required and "y" to represent the number of gallons of 65% solution.

Make an equation that represents gallons only: $x + y = 250$. This is the first equation.

Next, set up an equation representing the relative strengths of the solutions:

$$0.25x + 0.65y = 0.50 \bullet (250 gal.)$$

Solve these by adding the equations to make one of the variables disappear:

Multiply the second equation by -4:	$-4 \bullet (0.25x + 0.65y) = -4 \bullet (0.5 \bullet 250)$
Perform multiplication:	$-x - 2.6y = -500$
Add the first equation to the second:	$(x + y) - x - 2.6y = (250) - 500$
Combine like terms:	$-1.6y = -250$
Divide both sides by -1.6:	$y = 156.25$ gallons

Substitute this value in the first equation:	$x + 156.25 = 250$
Subtract 156.25 from both sides:	$x = 93.75$ gallons

Since x was the 25% solution, subtract the 93.75 gallons of x from the 250 gallons total to determine that y (the 65% solution) must be 156.25 gallons.

Check by entering 93.75 in the second equation:	$(0.25 \bullet 93.75) + (0.65 \bullet 156.25) = 0.5 \bullet 250$
Perform multiplication:	$23.438 + 101.562 = 125$
And perform addition:	$125 = 125$

Graphing, Slopes, and Intercepts

Equations with two variables are often graphed on an x-y axis to provide graphic representations of the data. Three-dimensional (3-D) graphs are also used to show more complex data, but 3-D graphs are not typically used on safety certification examinations. To graph data on an x-y axis, the x-value is plotted on a horizontal line (axis), and the y-value on a vertical one. Data values increase to the right on the x-axis and in the upward direction on the y-axis.

When neither the x-value nor the y-value is raised to any power other than one, the graph is a straight line. If the x or y variable has an exponent of two or greater, the graph will be a curved line or lines.

For straight lines, the "rise" divided by the "run" of such a line is called the *slope* of the line. Slopes can be positive or negative. When a slope is positive, the "y" variable increases when the "x" variable increases. When a slope is negative, the "y" variable decreases when the "x" variable increases.

To determine the slope of any line, any of the following formulas can be used:

Slope - intercept formula:	$y = mx + b$
Slope - x-intercept formula:	$y = m(x - a)$
Point - slope formula:	$y - y_1 = m(x - x_1)$
For all of the above formulas:	m = slope of the line
	x and y = coordinates of a known point on the line

APPLIED ALGEBRA

x_1, y_1 = coordinates of another known point on the line

a = the value of x when y is zero

b = the value of y when x is zero

None of the slope formulas are provided on certification exams. The certifying organizations consider these equations to be "relatively simple." You are expected to know enough to graph lines from formulas. Typically, knowing the first formula alone (slope – intercept formula) is adequate.

EXAMPLE

The graph of x = 2y would look like this:

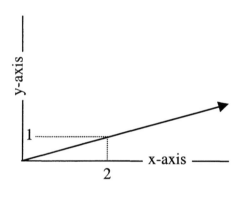

Given the formula x = 2y, for every increase of one unit in the y value, the x value increased by two. The slope of the equation can be determined by putting the equation into the slope-intercept form: $y = mx + b$. When you do this, the result is $y = \frac{x}{2} + 0$, which is the same as $y = 0.5x + 0$. This would mean that the slope of the line is positive one-half (0.5) and the b value (the value of y when x is zero) is zero also.

SETS, VENN DIAGRAMS, MIXTURE, AND GRAPHING QUESTIONS

a. Given sets A = {1,3,5,7,9} B = {1,3,5} and C = {1,2,3}, is set B a subset of A? Is set C a subset of A?

b. Given sets A = {1,3,5,7,9} and B = {1,2,3,4,5,6,7,8,9} draw a Venn diagram showing sets A & B.

c. 20 ounces of 65% nitric acid solution are needed for an experiment. The only solutions available are 50% and 75%. How much of each must be mixed to provide the proper amount and strength of nitric acid for the experiment?

CHAPTER 7

d. Given the equation $3x = y+2$, draw a graph of the line. What is the slope? What is the x-intercept? What is the y-intercept?

SETS, VENN DIAGRAMS, MIXTURE, AND GRAPHING ANSWERS

a. Yes, set B is a subset of A because each element of B is also an element of A.

No, set C is not a subset of A because one element of C (the two) is not an element of A.

b.
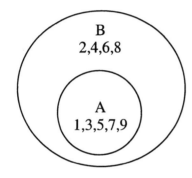

c. 8 ounces of 50% and 12 ounces of 75%, because:

Let "x" represent the ounces of 50% solution required. Let "y" represent the ounces of 75%.

Write the first equation (volume):	$x + y = 20$
Write the second equation (strength):	$0.5x + 0.75y = 0.65 \bullet 20$
Multiply:	$0.5x + 0.75y = 13$
Multiply the second equation by -2 so that "x" values will cancel out when the two equations are added together:	$-2 \bullet (0.5x + 0.75y) = -2 \bullet 13$
Perform the multiplication:	$-x - 1.5y = -26$
Add the first equation:	$-x - 1.5y + (x + y) = -26 + (20)$
Combine terms:	$-0.5y = -6$
And divide both sides by -0.5:	$y = 12$ ounces

110

Now substitute the 12 ounces
into the first equation: $x+12=20$

Subtract 12 from both sides: x = 8 ounces

Verify by substituting 8 into
the second equation: $(0.5 \bullet 8)+(0.75 \bullet 12)=0.65 \bullet 20$

Perform multiplication: $4+9=13$ or $13 = 13$

d. +3 = slope, because the slope is determined by using the slope-intercept formula ($y = mx + b$).

$3x = y+2$ becomes $y=3x-2$. Once in this form, the *m* value (the slope) is positive 3, and the *b* value (the value of *x* when *y* is zero) is negative two. This means that the line rises three units for every positive one unit of run.

-2 = y-intercept, because the y-intercept is determined by the *b* value in the above equation.

$\frac{2}{3}$ = x intercept, because the x-intercept can be determined by setting the *y* value to zero and solving for x: $3x=0+2$, so $x=\frac{2}{3}$

Graphed, the equation would look like this:

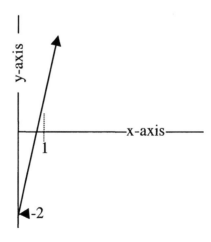

CHAPTER 7

Zero Product Property

Sometimes equations are given as several expressions multiplied together and set equal to zero. When this is the case the "zero product property" can apply. It says:

When multiplied expressions equal zero, one or more of the expressions must equal zero.

EXAMPLE

For $(2x+3)(x-7)=0$, either $(2x+3)$ must equal zero or $(x-7)$ must equal zero. To find the values for x, set the polynomials equal to zero and solve for x in both:

If $2x+3=0$, then $2x=-3$ and $x=-\frac{3}{2}$ or $x=-1.5$. If x is not -1.5, then:

$x-7=0$, so $x=7$. The only two valid answers for this expression (using the zero-product property) are $x = -1.5$ and/or 7.

Quadratic Equations

To be a quadratic equation, the following conditions must exist:

- The equation must have only one variable (no x-y or x-y-z equations).
- The highest order variable must be to the second power (squared).
- The equation must be written in the general form $ax^2 + bx + c = 0$ where a is not zero.

If these conditions are met, then the equation may be plugged into the quadratic formula (given in the *Candidates' Handbook* and on the formula sheet for examinations):

$$x = \frac{-b \pm \sqrt{b^2 - 4ac}}{2a}$$

Because of the plus or minus in the numerator, quadratic equations always have two solutions. You will need to examine the solutions in the context of the problem to determine which is most reasonable.

APPLIED ALGEBRA

EXAMPLE

Given the equation $x^2 + 3x + 2 = 0$, the equation is already in quadratic form. The constants can be plugged directly into the quadratic formula. Since there is only one x^2, the a value will be a positive one; the b value is a positive three; and the c value is a positive two.

$$x = \frac{-3 \pm \sqrt{3^2 - (4 \bullet 1 \bullet 2)}}{2 \bullet 1}$$

Simplify: $\quad x = \dfrac{-3 \pm \sqrt{9-8}}{2}$

Simplify again: $\quad x = \dfrac{-3 \pm \sqrt{1}}{2}$

This yields two answers: $\quad x = \dfrac{-3+1}{2}$ which is $\dfrac{-2}{2}$ or -1,

and $x = \dfrac{-3-1}{2}$ which is $\dfrac{-4}{2}$ or -2

Thus, $x =$ either -1 or -2 These solutions can be verified by plugging them back into the original equation.

If the same equation had been given in the form of $(x + 1)(x + 2) = 0$, the zero-product property could be used. In this form, either $x + 1$ must equal zero or $x + 2$ must equal zero. Solving both of these equations gives the same values for x (-1 or -2) as the quadratic equation.

Yet another way to solve quadratic equations on the exam is to try the given answer sets in the equation until you find the set that makes the equation true. Some calculators are preprogrammed with the quadratic equation. Others have programmable memories that allow you to enter and save the equation prior to the examination.

On safety certification examinations, quadratic equations are rarely given in the quadratic form (as in the example above). You are usually given a word problem, and expected to derive the equation prior to solving the quadratic.

CHAPTER 7

EXAMPLE

Four is added to the product of two positive, consecutive, even integers. The total is 84. What are the integers?

To solve this problem, an equation must be determined. If x is the smaller of the two consecutive even integers, the second is $x + 2$. When we add four to the product, we get 84. The equation looks like this:

$$[x \bullet (x+2)] + 4 = 84$$

Perform the multiplication: $\quad x^2 + 2x + 4 = 84$

Now subtract 84 from each side
to provide the general form: $\quad x^2 + 2x - 80 = 0$

Use the quadratic formula: $\quad x = \dfrac{-2 \pm \sqrt{2^2 - (4 \bullet 1 \bullet -80)}}{2 \bullet 1}$

Simplify: $\quad x = \dfrac{-2 \pm \sqrt{4 + 320}}{2}$ or $x = \dfrac{-2 \pm \sqrt{324}}{2}$

This yields two equations: $\quad x = \dfrac{-2 - 18}{2}$ and $x = \dfrac{-2 + 18}{2}$

These simplify to: $\quad x = \dfrac{-20}{2} = -10$ and $x = \dfrac{16}{2} = 8$

Both of these solutions are valid, but because the original problem specified positive integers, only the positive eight is an acceptable answer. Since the x was the lesser of the two consecutive, even integers, positive eight and ten are the answers.

Word Problems

Word problems are the most common way that math questions are presented on safety certification examinations. This is because word problems are also the most common way that "real-world" quandaries are presented to the working safety professional. The ability to derive solvable equations from word problems is the test of a safety professional's competence. There is no set method to apply to word problems, but the following sequence may be helpful:

First, read the problem carefully (or, in the real world, state the problem) to determine what is being asked for. Identifying the unknown element is crucial to this step. Don't try to solve the problem at this time.

Second, assign the unknown element a variable, such as x, and review the problem to see if all other unknown elements can be defined in terms of x. If not, select a different unknown element to be represented by x, and see if you can define other unknowns in terms of the new x. In some cases, a single variable is inadequate, and a second unknown, y, will have to be used. As a rule of thumb, the fewer the variables, the easier the equation is to solve—and the more likely you are to come up with the correct answer.

Third, review the entire problem, breaking it up into small pieces that can be represented by expressions. Watch for opportunities to set up ratios or proportions.

Fourth, combine the algebraic expressions from step three into a single equation (preferable) or multiple equations. Then solve the equations.

Finally, check the solution in the original equations and scenario to verify that it makes sense.

You are expected to know some background information in order to solve most word problems that you'll find on safety examinations. An example is *distance = velocity • time*. The *Candidates' Handbook* does not include this equation, since it's considered relatively simple. If the handbook did include this equation, it would be expressed as "$s = vt$," where s represents "displacement" or "distance," and v and t represent velocity and time, respectively. For example, if you travel at a velocity of 60 miles-per-hour for one hour's time, you have gone 60 miles. Use this equation in the following problem:

EXAMPLE 1

Two groups of students leave campus to go to a football game. The first car leaves on time, and travels at an average speed of 45 mph through the game-day traffic. The second car starts a half an hour later, but travels at a higher average speed of 55 mph. How long will it take for the second carload of students to catch up with the first?

To solve a problem of this kind, first read the problem to see what is asked for—in this case, the time it takes for the second car to catch up with the first, or t ("How long..."). Use a unit of "hours" since this makes sense (although minutes or even seconds could be used). Let the variable t equal the second car's time in hours.

Second, write one or more equations using the variable t and the $s = vt$ formula:

"Second car's distance equals the second car's velocity multiplied by the second car's time" becomes:

$\text{some distance} = 55 \bullet t$

"First car's distance equals the first car's velocity multiplied by the first car's time" becomes:

$\text{some distance} = 45 \bullet (t + 0.5)$

Since the first car traveled a half hour longer than the second, you add 0.5 (representing the extra half hour) to the second car's time, t.

Third, combine these equations into one, and then solve. Because the students started at the same place and ended at the same place, they must have traveled the same distance. This means that the *velocity • time* for the first car must equal *velocity • time* for the second. The equation may be stated as "first car's velocity multiplied by first car's time is equal to the second car's velocity multiplied by the second car's time":

$45 \bullet (t + 0.5) = 55 \bullet t$

Combine terms:	$45t + 22.5 = 55t$
Subtract 45x from each side:	$22.5 = 10t$
Divide both sides by 10:	$2.25 = t$

Since the variable t is in hours, the second car traveled two hours and fifteen minutes. (2.25 hours = 2 hours + 15 minutes)

To check this, plug this value back into either of the original equations:

First car's distance in miles:	$45 \bullet (2.25 + .05)$ or $45 \bullet 2.75 = 123.75$
Second car's distance in miles:	$55 \bullet 2.25 = 123.75$

Since the distances are identical (123.75 miles) the answer is valid.

EXAMPLE 2

Here's another example using simple mathematical terms: When three times the sum of four and an unknown number is subtracted from eight times the unknown number, the difference is equal to four times the sum of three and twice the unknown number. What is the unknown number?

First, read the problem carefully. It appears that one side of an equation is being set equal to a second set of terms.

Second, determine what is being asked for. In this case, it is simple—a single unknown number.

Third, write an equation based on the language of the problem:

"When three times the sum of four and an unknown number": $3 \bullet (4+x)$

"Is subtracted from eight times the unknown number": $8x - [3 \bullet (4+x)]$

"The difference is equal to four times the sum of three and twice the unknown number": $8x - [3 \bullet (4+x)] = 4 \bullet (3 + 2x)$

Simplify:	$8x - (12 + 3x) = 12 + 8x$
Combine terms:	$5x - 12 = 12 + 8x$
Subtract 5x from both sides:	$-12 = 12 + 3x$
Subtract 12 from both sides:	$-24 = 3x$
And divide both sides by 3:	$-8 = x$
To check, plug this value back into the original equation:	$[8 \bullet (-8)] - \{3 \bullet [4 + (-8)]\} = 4 \bullet \{3 + [2 \bullet (-8)]\}$
Simplify:	$(-64) - [3 \bullet (-4)] = 4 \bullet [3 + (-16)]$
Simplify again:	$-64 - (-12) = 4 \bullet (-13)$
And again:	$-64 + 12 = -52$
Finally:	$-52 = -52$

Since the equation balances, the answer is valid (assuming that you wrote the original equation correctly).

CHAPTER 7

EXAMPLE 3

Often, word problems can be solved most simply as ratios or proportions, assuming the problem can be structured that way. The following is an example of this.

If an investment of $7,000 earns $1,200 in 18 months, how much must be invested at the same rate to earn $1,500 in the same length of time?

First, read the problem to see what is given: The times for both investments are identical. The interest rates for both investments are identical. The only things that change are the starting investment amounts (in dollars) and the amount of interest earned (in dollars).

Second, determine what is being asked for-in this case, the starting amount ("how much") of the second investment amount must be determined. That amount (in dollars) will be x.

Third, write an equation showing the relationship between the first investment and the second. Since the interest rate and term are identical, the amount of yield earned by $7,000 over that time will be directly proportional to the amount of yield earned by x. Express this as a ratio: "The ratio of a starting amount of $7,000 to a yield of $1,200 is equal to the ratio of x to $1,500." Mathematically, this is expressed as:

$$\frac{\$7,000}{\$1,200} = \frac{x}{\$1,500}$$

Fourth, solve the equation. Once the equation is in the form of a proportion, cross-multiply to get:

$$1,200x = 7,000 \bullet 1,500$$

Simplify: $\quad\quad\quad\quad\quad\quad\quad\quad\quad 1,200x = 10,500,000$

Divide both sides by 1,200: $\quad\quad x = 8,750$

Review the original problem to verify that the answer makes sense. Since it took an original investment of $7,000 to earn a yield of $1,200, it should have taken a larger investment to earn a larger yield ($1,500). Since the answer, $8,750, is indeed larger than the original $7,000, the answer seems valid.

APPLIED ALGEBRA

EXAMPLE 4

Here's another example of proportions in word problems:

At 5 p.m., a 480-foot-tall building casts a shadow that is 1,248 feet long. A tree growing beside the building casts a shadow that is 140.4 feet long. How tall is the tree?

At first, this would seem to be a trigonometry problem, since triangles are involved, but it is actually a simple proportion problem. "The ratio of the height of the building to the length of its shadow equals the ratio of the height of the tree to the length of *its* shadow."

Express mathematically: $$\frac{480 \text{ ft.}}{1,248 \text{ ft.}} = \frac{x \text{ ft.}}{140.4 \text{ ft.}}$$

Cancel terms and cross multiply: $1,248x = 67,392$

Divide both sides by 1,248: $x = 54$ feet

Watch carefully for the opportunity to identify triangles of directly proportional shapes on safety certification examinations. Trigonometry can often be dispensed with under such circumstances.

EXAMPLE 5

Sometimes you are expected to know formulas that are not provided to solve word problems. An example:

Three laser printers can print out the same document in 20 minutes, 30 minutes, and 60 minutes respectively. How long would it take to finish the job if all three printers could simultaneously print parts of the same original document?

First, read the problem in depth. Obviously, if all three of the printers can simultaneously print parts of the same job, the overall time used must be less than that taken by the fastest (20 minute) printer alone.

Second, determine what is asked for. In this case, the time ("how long") is the variable of interest. Since all print times are given in minutes, it makes sense to define the variable t in minutes.

Third, write an equation that defines the problem. In this case, you must know that problems of this type just happen to use the same equation given in the

Candidates' Handbook for *either* resistors in parallel or capacitors in series. Define the equation to your needs as follows:

$\frac{1}{t_1}+\frac{1}{t_2}+\frac{1}{t_3}=\frac{1}{t_{tl}}$ where t is time, and the subscripts denote the individual printers' and the total times.

Fourth, plug in the times for the individual printers, and solve the equation:

$\frac{1}{20}+\frac{1}{30}+\frac{1}{60}=\frac{1}{t}$ Use your calculator to yield: $.0500+.0333...+.0166...=\frac{1}{t}$

This simplifies to: $0.1=\frac{1}{t}$ which is the same as $t=\frac{1}{0.1}$ or $t=10$ minutes.

Finally, check the answer to see if it makes sense. Since the time for all three printers running simultaneously on the single job must be less than the fastest printer by itself (20 minutes), the answer (10 minutes) seems plausible.

Another way to think of this problem is to say that in one minute, the 20-minute printer could print $\frac{1}{20}$ of the full job, the 30-minute printer could print $\frac{1}{30}$ of the full job, and the 60-minute printer could print $\frac{1}{60}$. Since they all run for the same amount of time in minutes (t minutes), the 20-minute printer completes $\frac{t}{20}$ of the full job, the 30-minute printer completes $\frac{t}{30}$, and the 60-minute printer $\frac{t}{60}$. The equation that results is: $\frac{t}{20}+\frac{t}{30}+\frac{t}{60}=1$. This simplifies to $0.1t=1$ or $t=10$ minutes.

APPLIED ALGEBRA

ZERO-PRODUCT, QUADRATIC EQUATION, PROPORTIONAL TRIANGLE, AND WORD PROBLEM QUESTIONS

a. Solve the following equation using the zero-product property: $(4x+8) \bullet (x-3) = 0$

b. $x^2 - 4x + 3 = 0$ What is x?

c. A worker can ride a bicycle at a rate that is five miles per hour faster than she can walk. If the worker walks for two hours and then bicycles for three hours, she covers a distance of 35 miles. How fast does she walk?

d. When two times the sum of four plus an unknown number is subtracted from eight times the unknown number, the difference is equal to the unknown number plus the sum of three and two times the unknown number. What is the unknown number?

e. A chain of retail stores is growing in a metropolitan area. The owner currently has four stores, and reaps a total annual profit of $256,000. The owner wishes to increase profits to an annual total of $448,000. How many additional stores must be opened?

f. Three valves of various sizes, working together, can drain a dike in twelve hours. If one valve can empty the dike by itself in eighteen hours, how much time will it take for the second and third valves (working together without the first valve) to drain the dike?

g. A stadium arc light is mounted 50 feet above a playing field. The wall that surrounds the playing field is 75 feet in front of the arc light pole. If the field wall is 4 feet tall, how long is the shadow it casts?

ZERO-PRODUCT, QUADRATIC EQUATION, PROPORTIONAL TRIANGLE, AND WORD PROBLEM ANSWERS

a. -2 or 3, because one or both of the two expressions (4x + 8 or x - 3) must equal zero. Therefore:

$4x + 8 = 0$ is the same as $4x = -8$ which simplifies to $x = \dfrac{-8}{4}$ or $x = -2$

$x - 3 = 0$ is the same as $x = 3$

b. 1 or 3. This can be determined several ways. Using the zero-product property, you can factor the equation to: $(x-1)(x-3)=0$ which yields either $x-1=0$ and/or $x-3=0$. Another option is to use the quadratic formula given in the *Candidates' Handbook*: $x=\dfrac{-b\pm\sqrt{b^2-4ac}}{2a}$.

This yields: $x=\dfrac{4\pm\sqrt{16-(4\bullet 3)}}{2}$ which is the same as $x=\dfrac{4\pm 2}{2}$

Condition one yields: $x=\dfrac{4+2}{2}=\dfrac{6}{2}=3$

Condition two yields: $x=\dfrac{4-2}{2}=\dfrac{2}{2}=1$

c. 4 miles per hour. To solve this, use the "rate-times-time-equals-distance" formula:

Let the walking rate (in miles-per-hour) equal x.

$2\bullet x+[3\bullet(x+5)]=35$

Simplify: $\qquad 2x+3x+15=35$

Combine terms: $\qquad 5x+15=35$

Subtract 15 from each side: $\quad 5x=20$

And divide both sides by 5: $\quad x=4$ mph

d. Three and two thirds is the unknown number because:

$8x-2(4+x)=x+(3+2x)$

Simplify: $\qquad 8x-8-2x=3x+3$

Combine terms: $\qquad 6x-8=3x+3$

Subtract 3x from both sides: $\quad 3x-8=3$

Add 8 to both sides: $\qquad 3x=11$

And divide both sides by 3: $\quad x=\dfrac{11}{3}=3\tfrac{2}{3}$

e. Three. Assume that net profit per store is constant and set up a direct proportion phrased as follows: "The ratio of four stores to a $256,000 annual profit equals the ratio of x stores to a $448,000 annual profit."

APPLIED ALGEBRA

Express mathematically: $\dfrac{4}{256{,}000} = \dfrac{x}{448{,}000}$

Cross-multiply: $256{,}000x = 1{,}792{,}000$

Divide both sides by 256,000: $x = 7$.

Since it took seven stores to produce the desired profit, and the owner already had four, 7 − 4 = 3. The owner needs three new stores to attain the goal.

f. 36 hours, because x is the time it takes the second and third valves *combined* to drain the dike:

$$\dfrac{1}{18} + \dfrac{1}{x} = \dfrac{1}{12}$$

Find a common denominator for the left side: $\left(\dfrac{x}{x} \bullet \dfrac{1}{18}\right) + \left(\dfrac{18}{18} \bullet \dfrac{1}{x}\right) = \dfrac{1}{12}$

Simplify: $\dfrac{x}{18x} + \dfrac{18}{18x} = \dfrac{1}{12}$

Combine terms: $\dfrac{x+18}{18x} = \dfrac{1}{12}$

Cross-multiply: $12x + 216 = 18x$

Subtract $12x$ from both sides: $216 = 6x$

And divide both sides by 6: $36 = x$

g. Approximately 6.52 feet, because:

First, draw the problem to show the relationships:

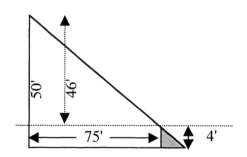

CHAPTER 7

The larger triangle is directly proportional to the small one defined by the field wall and its shadow. The "larger" triangle you will use is the one defined by the shadowed lines, since you know the lengths of two of its three sides. The base of the larger triangle is 75 feet (the distance from the light pole to the field wall). The height of the larger triangle is 46 feet (the distance from the top of the field wall to the top of the light pole). The height of the small triangle will be four feet (the height of the field wall). Its base is the shadow length that you wish to find. Because the triangles are directly proportional, you can say that "the ratio of the height of the larger (L) triangle to its base is equal to ratio of the height of the smaller (S) triangle to *its* base."

Express mathematically: $$\frac{"L" \text{ height}}{"L" \text{ base}} = \frac{"S" \text{ height}}{"S" \text{ base}}$$

Enter the data: $$\frac{46}{75} = \frac{4}{x}$$

Cross multiply: $300 = 46x$

And divide both sides by 46: $6.52 = x$ feet (approximately)

APPLIED ALGEBRA CHALLENGE EXAM

a. Given set A = {1,3,5,7,9}, is 4 an element of A? Is set B = {1,3,7} a subset of set A? Is set C = {1,2,3} a subset of A?

b. Given sets A = {1,3,5,7,9}, set B = {1,2,3,4,5}, and set C = {5,7,9}, draw a Venn diagram showing the relationships.

c. Given the equation $y = x^2 + 3$, will the graph yield a straight line?

d. Given the equation $4 = 3y - 2x$, what is the slope? What is the x-intercept? What is the y-intercept?

e. Given the equation $(3x-5)(x+4)(2x-2) = 0$ solve for x.

f. Given the equation $28 - 3x = 4x(x-5)$, solve for x.

g. It takes an industrial forklift three fewer minutes than a small front-end-loader to traverse the length of a warehouse. Find the length of the warehouse if the front-end-loader can travel at three mph and the forklift at four.

h. In a two-digit number, the units digit that is one less than the tens digit. If the product of digits is divided by their sum, the answer is one and one fifth. What is the two digit number?

i. A reactor has two fill pipes and a drain. The first fill pipe charges the reactor in twelve minutes. The second fill pipe can charge the reactor in four minutes. When the drain is opened, it can drain the reactor in six minutes. How long does it take to charge the reactor if both fill lines and the drain are simultaneously opened?

j. Compressor A takes 36 minutes to pump the same volume of gas that compressor B can pump in 24 minutes. If compressor A is started six minutes before compressor B, how long does it take for the pair of them to pump the equivalent volume that A or B can pump in 36 minutes or 24 minutes, respectively?

APPLIED ALGEBRA CHALLENGE EXAM ANSWERS

a. 4 is not an element of set A because set A does not contain a 4.

Set B is a subset of A because all elements of B (1,3,7) are also elements of A.

Set C is not a subset of A because one element of C (the 2) is not an element of A.

b. The Venn diagram of these sets looks as follows:

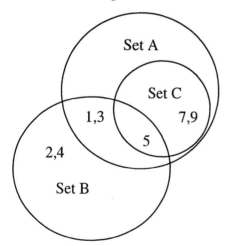

c. No. Only equations with x and y values to the power of one yield straight lines when graphed.

d. Slope = 2/3, y-intercept = 4/3, and x-intercept = -2 Use the original equation (4 = 3y-2x)

Write the equation in slope-intercept form: $(y = mx + b)$

Add 2x to both sides of the equation: $3y = 2x + 4$

Divide both sides of the equation by three: $y = \dfrac{2}{3}x + \dfrac{4}{3}$

This tells you that the slope of the line is 2/3 (the *m* value) and the y-intercept is 4/3 (the *b* value)

To calculate the x-intercept, set the y value to zero in the original equation and solve for x:

$4 = 3 \bullet 0 - 2x$ becomes 4 = -2x, which becomes -2 = x.

e. $x = \dfrac{5}{3}$, -4, or 1, because given the original equation $(3x-5)(x+4)(2x-2) = 0$, the zero-product property states that one or more of the three polynomials must equal zero.

If $3x - 5 = 0$, then

    Add 5 to both sides: $3x = 5$

    Divide both sides by 3: $x = \dfrac{5}{3}$

If $x + 4 = 0$, then $x = -4$

If $2x - 2 = 0$, then

    Add 2 to both sides: $2x = 2$

    Divide both sides by 2: $x = \dfrac{2}{2}$, or $x = 1$

f. $x = -1.268$ or 5.518, because the equation simplifies to: $28 - 3x = 4x(x-5)$

Perform the multiplication: $28 - 3x = 4x^2 - 20x$

Subtract $4x^2$ from both sides: $-4x^2 - 3x + 28 = -20x$

Add 20x to both sides: $-4x^2 - 3x + 20x + 28 = 0$

APPLIED ALGEBRA

Combine terms:	$-4x^2 + 17x + 28 = 0$
Apply the quadratic equation:	$x = \dfrac{-b \pm \sqrt{b^2 - 4ac}}{2a}$ becomes
	$x = \dfrac{-17 \pm \sqrt{17^2 - 4 \bullet (-4) \bullet 28}}{2 \bullet (-4)}$
Combine terms:	$x = \dfrac{-17 \pm \sqrt{289 - (-448)}}{-8}$
Combine terms again:	$x = \dfrac{-17 \pm \sqrt{737}}{-8}$
The answers are:	$x = \dfrac{-17 + 27.148}{-8}$ and $x = \dfrac{-17 - 27.148}{-8}$
For the first case:	$x = \dfrac{10.148}{-8}$ or $x = -1.268$
For the second:	$x = \dfrac{-44.148}{-8}$ or $x = 5.518$

g. 0.6 miles Because using "rate times time equals distance," you have:

$3 mph \bullet t =$ warehouse length and $4 mph \bullet (t-3) =$ warehouse length

So:	$3t = 4(t-3)$ Where t is a time in minutes
Perform the multiplication:	$3t = 4t - 12$
Subtract $4t$ from both sides:	$-t = -12$
Multiply both sides by negative one:	$t = 12$ minutes
	(to traverse the warehouse length)

If the front-end-loader travels 3 miles in 60 minutes, then 3 miles is to 60 minutes as x miles is to twelve-minutes:

$$\dfrac{3 mi.}{60 min.} = \dfrac{x}{12 min.}$$

Cross multiply:	$60 min. \bullet x = 36 \bullet mi. \bullet min.$
And divide by 60 minutes:	$x = 0.6$ miles

h. 32 Let x equal the tens digit, and $x - 1$ must equal the units digit.

Read the problem and
write the equation: $\dfrac{x(x-1)}{x+(x-1)} = \dfrac{6}{5}$

Perform multiplication: $\dfrac{x^2 - x}{2x - 1} = \dfrac{6}{5}$

Cross multiply: $5x^2 - 5x = 12x - 6$

Subtract $12x$ from both sides: $5x^2 - 17x = -6$

Add 6 to both sides: $5x^2 - 12x + 6 = 0$

Use the quadratic equation: $x = \dfrac{-b \pm \sqrt{b^2 - 4ac}}{2a}$

$x = \dfrac{-(-17) \pm \sqrt{(-17)^2 - (4 \bullet 5 \bullet 6)}}{2 \bullet 5}$

Perform multiplication: $x = \dfrac{17 \pm \sqrt{289 - 120}}{10}$

Simplify: $x = \dfrac{17 \pm \sqrt{169}}{10}$

This yields two equations: $x = \dfrac{17 + 13}{10}$ and $x = \dfrac{17 - 13}{10}$

Solve: $x = \dfrac{30}{10} = 3$ and $x = \dfrac{4}{10} = 0.4$

Since you must have an integer, only the $x = 3$ makes sense. Since x stood for the tens digit, the units digit ($x-1$) is 2, and the number is 32.

i. 6 minutes. In this case, x represents time (in minutes) to fill the reactor with all three lines open. Because of this:

Fill pipe one's rate is 12 minutes, fill pipe two's rate is 4 minutes, and the drain line's rate is 6 minutes.

APPLIED ALGEBRA

Therefore:
$$\frac{1}{12}+\frac{1}{4}-\frac{1}{6}=\frac{1}{x}$$

(Note that the drain line was subtracted from the total, and that the x on the right of the equation represents total time in minutes.)

Find a common denominator:
$$\left(\frac{1}{12}\right)+\left(\frac{3}{3}\cdot\frac{1}{4}\right)-\left(\frac{2}{2}\cdot\frac{1}{6}\right)=\frac{1}{x}$$

Perform multiplication:
$$\frac{1}{12}+\frac{3}{12}-\frac{2}{12}=\frac{1}{x} \text{ becomes } \frac{1+3-2}{12}=\frac{1}{x}$$

Combine terms:
$$\frac{2}{12}=\frac{1}{x}$$

Cross multiply: $2x = 12$

Divide both sides by 2: $x = 6$ minutes

j. 18 minutes. Note that another variable has been added to this problem. The compressors not only pump at different rates, but they run for different amounts of time. Using a "rate times time equals volume" equation, let x equal the combined time (in minutes) needed to pump the amount of gas, given the problem's conditions:

Compressor A's time = x minutes

Compressor A's rate is 1 / 36 amount per minute

Compressor B's time = $(x - 6)$ minutes

Compressor B's rate is 1 / 24 amount per minute

Volume of gas we seek = 1 amount. This requires some explanation:

Because the two compressors do not end up pumping the same volume, we can't set the "rates-times-times" equal to each other for this equation. Instead, the total volume of gas that we seek is calculated by multiplying either compressor's rate by the time that that compressor (alone) would have to run to pump the volume we seek. For compressor A, that would be one-thirty-sixth of the amount-per-minute times 36 minutes, which equals one. The number one represents the total volume of gas we seek.

For this problem, x represented the minutes that compressor A ran. Therefore, compressor A's "rate-times x" added to compressor B's "rate times $(x$ minus 6-minutes)" must equal one volume of gas.

In equation form:	$\dfrac{1}{36}x + \dfrac{1}{24}(x-6) = 1$
Find a common denominator:	$\left(\dfrac{24}{24} \cdot \dfrac{1}{36} \cdot \dfrac{x}{1}\right) + \left(\dfrac{36}{36} \cdot \dfrac{1}{24} \cdot \dfrac{(x-6)}{1}\right) = 1$
Perform multiplication:	$\dfrac{24x}{864} + \dfrac{36x - 216}{864} = 1$ becomes $\dfrac{24x + 36x - 216}{864} = 1$
Combine terms:	$\dfrac{60x - 216}{864} = 1$
Multiply both sides by 864:	$60x - 216 = 864$
Add 216 to both sides:	$60x = 1{,}080$
Divide both sides by 60:	$x = 18$ minutes

CHAPTER EIGHT

Geometry

> **Assumptions:** You have sufficient knowledge to take and pass the challenge exam of all previous chapters. You have a scientific calculator and basic knowledge of how to use it.
>
> **Application:** Required for ASP, CSP, OHST, and CHST examinations.
>
> **Discussion:** Understanding geometry is essential, because many real-world problems require the ability to evaluate shapes for their area or volume. The ability to use geometry to calculate area and volume is a skill that will be required consistently in both safety certification examinations and in real life. If you feel that your comprehension of geometry is good, go to the challenge exam at the end of this chapter. If you can work all problems correctly, skip this chapter.

Geometry is defined as the study of the properties and relationships of points, lines, angles, and shapes. For the purposes of certification examinations, the main geometric definitions of interest are:

Perimeter: The total of the *lengths* of the edges of a shape; in other words, the distance around the shape.

Circumference: The perimeter of a circle

Diameter: The distance across a circle, through its center, also equal to the circle's width.

Radius: The distance from the center of a circle to any point on the perimeter, also one-half the circle's width.

Area: A two-dimensional measurement of the space contained by a shape.

Volume: A three-dimensional measurement of the space contained by a shape.

The above definitions and the following formulas are not given in *Candidates' Handbook*. They are considered relatively simple, and you will be expected to know them. The small square boxes in corners of shapes indicate 90-degree (right) angles. You should know how to calculate the perimeter, circumference, diameter, radius, area, and/or volume of the following common shapes:

CHAPTER 8

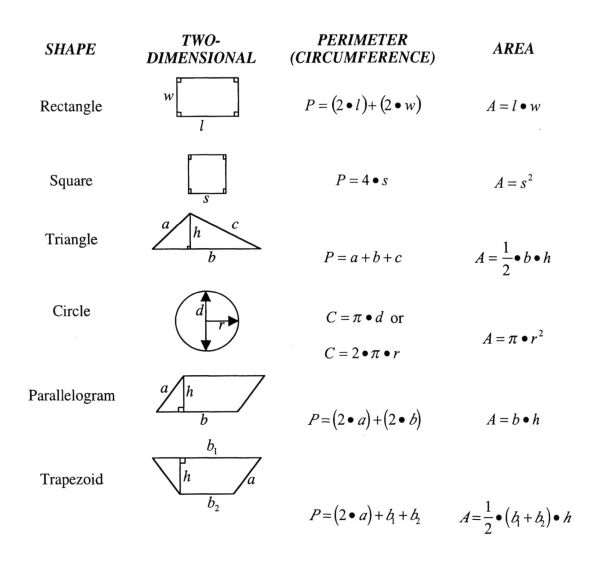

Table 12: Two-Dimensional Shapes

GEOMETRY

FIGURE	THREE-DIMENSIONAL	VOLUME
Rectangular Solid	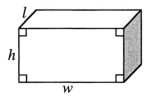	$V = l \bullet w \bullet h$
Cube	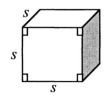	$V = s^3$
Cylinder	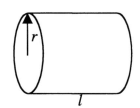	$V = \pi \bullet r^2 \bullet l$
Cone		$V = \frac{1}{3} \bullet \pi \bullet r^2 \bullet h$
Sphere	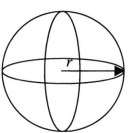	$V = \frac{4}{3} \bullet \pi \bullet r^3$

Table 13: Three-Dimensional Shapes

CHAPTER 8

Geometry is often used on certification examinations in "compound" problems that require you to know both the geometry formula and additional information. The following examples illustrate compound problems using geometry:

EXAMPLE 1

The length of a living room is one foot more than twice the width. The perimeter is 110 feet. Find the length and width of the room.

Since both the length and width are asked for, you can arbitrarily assign either one to be the x value in this problem. If "width" is the x-variable, then the length will be one foot more than twice x, or $length = 2x+1$. Since the perimeter of a rectangle is specified by the equation "$P = (2 \bullet l) + (2 \bullet w)$," substitute the length and width values as follows: $110 = [2 \bullet (2x+1)] + (2 \bullet x)$

Multiply:	$110 = (4x+2) + (2x)$
Combine terms:	$110 = 6x + 2$
Subtract two from each side:	$108 = 6x$
And divide both sides by 6:	$18 = x$

Because x was the room's width in feet, use the $length = 2x+1$ formula to determine the length. $length = (2 \bullet 18) + 1$, or 37 feet.

EXAMPLE 2

A 50' x 150' flat roof has a three-inch accumulation of standing water. What is the weight of the water?

Weight is the unknown variable in this problem. Before weight can be calculated, though, the volume of water must be determined (using geometry). Since the *Candidates' Handbook* gives the weight of water as 62.4 pounds per cubic foot, it makes sense to calculate the volume in cubic feet as well.

Since the volume of a rectangular solid is:	$V = l \bullet w \bullet h$
Enter all dimensions in feet:	$V = 150\,ft. \bullet 50\,ft. \bullet \left(\dfrac{3in.}{1} \bullet \dfrac{1ft.}{12in.} \right)$
	(note the conversion of 3" to ft.)
Perform multiplication:	$Volume = 1,875\,ft.^3$

GEOMETRY

Since each cubic foot of water weighs 62.4 pounds:

$$\frac{1{,}875\,ft^3}{1} \cdot \frac{62.4\,lb.}{1\,ft^3} = 117{,}000 \text{ pounds}$$

EXAMPLE 3

A conical pile of wood chips is 150 feet high. The base is twice as wide as the height. A conveyer belt removes chips from the pile at a rate of two cubic feet per second. If the conveyer belt operates continuously, how many hours will it take to consume the pile of chips?

In this problem, the conveyer belt's rate is given in cubic feet per second, but the answer must be in hours. One formula used for this problem is *Volume = Rate • Time*. Set the total volume of the chip pile equal to the rate of the conveyer belt (given as 2-cubic-feet-per-second) multiplied by the time taken to move the whole pile (the unknown). Before the problem can be solved, the volume of the wood-chip pile must be determined using geometry:

The formula for the volume of a cone is:

$$V = \frac{1}{3} \cdot \pi \cdot r^2 \cdot h$$

The base diameter is 300' (2 • height), so the radius is 150':

$$V = \frac{1}{3} \cdot \pi \cdot (150\,ft.)^2 \cdot 150\,ft.$$

Perform multiplication:

$$V = 3{,}534{,}292\,ft^3, \text{ approximately}$$

Dividing by the rate (cfs) of the conveyer:

$$\frac{3{,}534{,}292\,ft^3}{\frac{2\,ft^3}{1\,sec.}}$$

Invert and multiply:

$$\frac{3{,}534{,}292\,ft^3}{1} \cdot \frac{1\,sec.}{2\,ft^3} = 1{,}767{,}146\,sec.$$

Convert to hours:

$$\frac{1{,}767{,}146\,sec.}{1} \cdot \frac{1\,min.}{60\,sec.} \cdot \frac{1\,hour}{60\,min.} = 491 \text{ hours}$$

CHAPTER 8

GEOMETRY QUESTIONS

a. If a given circle has a diameter of ten feet, what is its circumference? What is its area?

b. What is the area of the triangle at right?

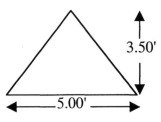

c. A trapezoidal piece of property is to be purchased that has dimensions of three miles along one side and two miles along the parallel side. The parallel sides are one mile apart. What is the area of the property?

d. A pipe full of methanol is 300 yards long and has an internal diameter of four inches. How many cubic feet of methanol does the pipe contain?

e. A liquefied-gas storage sphere has a circumference of 314.2 feet. What is its volume in cubic feet?

f. A dip pan is to be fabricated from a 12-foot-square piece of sheet metal by removing a two-foot square from each corner and folding. What is the capacity of the dip pan in cubic meters?

g. A single-piston reciprocating pump has a 15 cm diameter piston and a 15 cm stroke. If the pump makes 16 strokes per minute, what is the flow rate in gpm?

GEOMETRY ANSWERS

a. Approximately 31 feet and 78 square feet, because:

The formula for circumference is $C = \pi \bullet d$; therefore for a ten-foot circle, $C = 3.142 \bullet 10$, or 31.42 feet. This is the distance around the edge of the circle.

The formula for area of a circle is $A = \pi \bullet r^2$; therefore for a ten-foot circle, the radius is five feet.

$A = \pi \bullet 5^2$, or $\pi \bullet 25 = 78.54$ square feet.

b. 8.75 square feet, because the formula for area of a triangle is: $A = \frac{1}{2} \bullet b \bullet h$. This triangle's base is 5 feet and the height 3.5 feet. $A = \frac{1}{2} \bullet 5 \bullet 3.5 = 8.75$. Because all original measurements were carried to two decimal places, the correct answer is 8.75 square feet.

c. 2.5 square miles, because the formula for the area of a trapezoid is $A = \frac{1}{2} \bullet (b_1 + b_2) \bullet h$. The b measurements represent the parallel sides, and the h represents the distance between the parallel sides. Substituting, the formula becomes: $A = 0.5 \bullet (3+2) \bullet 1$, which simplifies to: $A = 0.5 \bullet 5$, or 2.5. Since the units were miles, the correct answer is 2.5 square miles.

d. Approximately 78.54 cubic feet. Since the pipe in the problem is assumed to be round, the pipe is a cylinder for purposes of the problem. The formula for volume of a cylinder is $V = \pi \bullet r^2 \bullet l$. Note that this problem (as is typical on certification examinations) gives the problem in differing units (the dimensions are given in inches and yards, but the problem requires an answer in cubic feet). Before the problem can be worked, all dimensions must be converted to identical units. Because "cubic-feet" is the unit requested as an answer, convert all dimensions to feet first:

Cylinder length (l) becomes $\frac{300 \, yd.}{1} \bullet \frac{3 \, ft.}{1 \, yd.} = 900 \, ft.$

Cylinder radius (r) goes from 2" (half of the 4" diameter given) to: $\frac{2 \, in.}{1} \bullet \frac{1 \, ft.}{12 \, in.} = 0.166 \, ft.$

Finally, using the formula $V = \pi \bullet 0.166..^2 \bullet 900$ gives an answer of approximately $78.54 \, ft.^3$

e. Approximately 523,599 cubic feet. Because the problem provides the circumference of the sphere, work backward to get the diameter, and thus the radius needed to calculate the volume. The formula for the circumference of a circle is: $C = \pi \bullet d$. Use algebra to divide both sides by pi, yielding $d = \frac{C}{\pi}$. Perform the division: $d = \frac{314.2}{\pi}$, or d = 100. Since the circumference was given in feet, the diameter of the sphere is 100 feet. The radius (half the diameter), is therefore fifty feet. Use the formula for the volume of a sphere:

$V = \frac{4}{3} \bullet \pi \bullet r^3$.

Enter the data: $V = \dfrac{4}{3} \bullet \pi \bullet 50^3$

Perform multiplication: $V = \dfrac{4}{3} \bullet \pi \bullet 125{,}000$ which is about 523,599 cubic feet.

f. Approximately 3.625 cubic meters First, determine the capacity of the dip pan. It is sometimes helpful to draw such problems in order to visualize them:

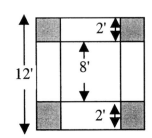

Since the overall dimension of the piece of sheet metal are 12' x 12', the "sides" of the pan are 2' tall. This means that the center square, or "bottom" of the pan must be 8' x 8'. Use the formula for the volume of a rectangular solid:
$V = l \bullet w \bullet h$, and plug in the length and width as 8' each (the dimensions of the pan's bottom), and the height as 2'. Then convert from cubic feet to cubic meters to answer the problem.

Volume of the pan
as constructed: $\quad V = 8 \bullet 8 \bullet 2 = 128\,ft^3$

Using the conversion from
the *Candidates' Handbook*: $\quad \dfrac{128\,ft^3}{1} \bullet \dfrac{1\,liter}{0.03531\,ft^3} \bullet \dfrac{1\,m^3}{1{,}000\,liters} = 3.625$ cubic meters

g. Approximately 11.24 gpm, because:

The volume the piston (a cylinder): $V = \pi \bullet r^2 \bullet l$ becomes $V = \pi \bullet 7.5^2 \bullet 15$

Perform multiplication: $\quad V = 2{,}650.72\,cm^3$

Since the pump makes 16 strokes/min: $\quad \dfrac{2{,}650.72\,cc}{1\,stroke} \bullet \dfrac{16\,strokes}{1\,min.} = \dfrac{42{,}412\,cc}{1\,min.}$

Convert from cc to gallons: $\quad \dfrac{42{,}412\,cc}{1\,min.} \bullet \dfrac{1\,liter}{1{,}000\,cc} \bullet \dfrac{1.06\,qt.}{1\,liter} \bullet \dfrac{1\,gal.}{4\,qt.} = \dfrac{11.24\,gal.}{1\,min.}$

GEOMETRY

GEOMETRY CHALLENGE EXAM

a. Debris is thrown from the surface of a grinding wheel at a velocity of 2,617 fpm. What is the diameter of the grinding wheel in mm, assuming 1,000 rpm?

b. A pipeline segment is to be cleared for maintenance work. The internal diameter of the pipe section is 9 inches. The section is twelve feet long. How many gallons of catch capacity are required to drain the liquid contained in this section when the flange is broken?

c. A pit in the ground that contains telephone switching equipment measures five feet wide by ten feet long by seven feet deep. A walkway extends completely around the top of the pit. The walkway is one foot narrower on the ends of the pit than on the (longer) sides. The total area of the walkway is 104 square feet. How wide is the walkway on the ends and on the sides?

d. A round duct has an internal radius of eight inches. Air flows through the duct at a velocity of 100fpm. What volume (cfm) of air is moved?

e. What is the perimeter of an equilateral triangle with sides of ten meters?

f. A hundred lengths of pipe must be delivered to a construction site. The pipes are each seven feet long, and have an outside diameter of 2.2 inches with an inside diameter of 2 inches. The piping material weighs 0.3 pounds per cubic inch. Will the hundred pipes overload a pickup truck with a one-ton capacity?

g. A cylinder has an inner diameter of seven feet and a height of twenty feet. Inside the cylinder are three spherical reactors, each of which has an outer diameter of six feet. Ignoring connecting piping and support brackets, how much of the space in the cylinder is not occupied by the reactors?

GEOMETRY CHALLENGE EXAM ANSWERS

a. Approximately 254mm. Because debris is thrown at the surface speed of a moving surface, the grinding wheel's edge must be moving at 2,617 feet-per-minute.

Determine the circumference of the wheel:
$$\frac{2,617\,ft.}{1\,min.} \bullet \frac{1\,min.}{1,000\,rev.} = \frac{2.617\,ft.}{1\,rev.} = C$$

Determine the wheel's diameter: $C = \pi \bullet d$ becomes $2.617\,ft. = 3.142 \bullet d$

Divide both sides by pi: $0.833\,ft. = d$

Convert to mm:

$$\frac{0.833\,ft.}{1} \bullet \frac{12in.}{1ft.} \bullet \frac{2.54cm}{1in.} \bullet \frac{10mm}{1cm} = 254mm$$

b. Approximately 39.78 gallons. Since the *Candidates' Handbook* gives the conversion factor of 1.06 qt. = 0.03531 cubic feet, you will calculate the volume of the pipe in cubic feet:

First, convert all units to feet: $\quad \dfrac{9in.}{1} \bullet \dfrac{1ft.}{12in.} = 0.75\,ft.$ diameter of pipe

Volume of a cylinder: $\quad V = \pi \bullet r^2 \bullet l \text{ becomes } V = \pi \bullet \left(\dfrac{0.75\,ft.}{2}\right)^2 \bullet 12\,ft.$

Perform multiplication: $\quad V = 5.301\,ft.^3$

Convert to gallons: $\quad \dfrac{5.301\,ft.^3}{1} \bullet \dfrac{1.06qt.}{0.03531\,ft.^3} \bullet \dfrac{1gal.}{4qt.} = 39.78$ gallons.

c. 2 feet wide on the ends of the pit, and three feet wide along the (longer) sides.

In this problem, it is helpful to draw the pit to help visualize the walkway:

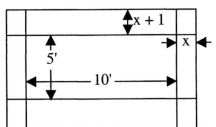

Since the area of a rectangle is determined by the formula $A = l \bullet w$, the area of the walkway alone will be the area of the overall shape minus the area of the central rectangle (pit).

The overall length will be ten feet plus the "x-width" walkways on both ends. The depth of the pit is extra information that is not needed to solve the problem. Express the length algebraically: $10 + 2x = l$.

The overall width will be five feet plus the "x-plus one-foot-width" walkways on either side. Expressed algebraically: $5 + 2(x+1) = 5 + 2x + 2 = 7 + 2x$. Now, set the total area of the shape equal to the area of the walkway $\left(104\,ft.^2\right)$ plus the area of the center pit $(5 \bullet 10) = 50\,ft.^2$ Expressed algebraically, the equation becomes:
$(7 + 2x) \bullet (10 + 2x) = 104 + 50$

Perform multiplication:	$70 + 14x + 20x + 4x^2 = 154$
Combine terms:	$4x^2 + 34x + 70 = 154$
Subtract 154 from both sides:	$4x^2 + 34x - 84 = 0$
Use the quadratic formula:	$x = \dfrac{-b \pm \sqrt{b^2 - 4ac}}{2a}$
This becomes:	$x = \dfrac{-34 \pm \sqrt{34^2 - 4 \cdot 4 \cdot (-84)}}{2 \cdot 4}$
Perform multiplication:	$x = \dfrac{-34 \pm \sqrt{1{,}156 + 1{,}344}}{8}$
Simplify:	$x = \dfrac{-34 \pm \sqrt{2{,}500}}{8}$ or $x = \dfrac{-34 \pm 50}{8}$
This yields:	$x = \dfrac{-34 + 50}{8} = \dfrac{16}{8} = 2$ and
	$x = \dfrac{-34 - 50}{8} = \dfrac{-84}{8} = -10.5$

Since -10.5 feet makes no sense as a walkway width, the ends of the tank have a two-foot-wide walkway, and the sides have (*x* + 1) a three-foot-wide one.

d. Approximately 139.626 cubic feet per minute. The *Candidates' Handbook* gives this formula in the "ventilation" section: Q = AV, where Q = Volume (in cubic units), A = Cross-sectional area of duct, pipe, or shape, and V = Velocity of gas in linear units per time. Note that all units must match! Don't mix inches, feet, and meters. To set up this problem, first find A (area of the duct). Since the velocity is given in feet per minute, we will convert the duct's area to feet as well:

First convert the internal radius to feet: $\dfrac{8 in.}{1} \cdot \dfrac{1 ft.}{12 in.} = 0.667 \, ft.$

Second, calculate the area of the duct: $A = \pi \cdot (0.667 \, ft.)^2 = 1.396 \, ft^2$

Third, calculate the volume using Q=AV:

$$Q = \left(\frac{1.396 ft.^2}{1}\right) \bullet \left(\frac{100 ft.}{1 min.}\right) = \frac{139.626 ft.^3}{1 min.}$$

e. 30 meters, because each of the three sides is ten meters long; add the three sides to calculate the perimeter.

f. No, the load (approximately 1,661 pounds) is less than the one-ton (2,000 pound) capacity of the truck.

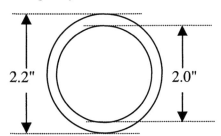

Area of the pipe outside diameter = $\pi \bullet \left(\frac{2.2 in.}{2}\right)^2$

Area of the pipe inside diameter = $\pi \bullet \left(\frac{2.0 in.}{2}\right)^2$

Working the math, OD = 3.801 square inches.

Pipe ID = 3.142 square inches.

Pipe OD minus pipe ID yields area of the pipe cross section (wall material only).

3.801 (OD) minus 3.142 (ID) = 0.659 square inches.

Multiply by the pipe's length: $\frac{0.659 in.^2}{1} \bullet \left(\frac{7 ft.}{1} \bullet \frac{12 in.}{1 ft.}\right) = 55.356 in.^3$

Since each cubic inch weighs 0.3#: $\frac{55.356 in.^3}{1} \bullet \frac{0.3 lb.}{1 in.^3} = 16.607$ pounds per pipe.

Because there are a hundred pipes: $16.607 \bullet 100 = 1,660.7$ pounds.

g. Approximately 431 cubic feet.

Calculate the volume of the cylinder: $\pi \cdot \left(\dfrac{7\,ft.}{2}\right)^2 \cdot 20\,ft.$

Simplify: $\pi \cdot 3.5^2 \cdot 20$ or $770\,ft.^3$ (approximately)

Calculate the volume of one sphere: $\dfrac{4}{3} \cdot \pi \cdot \left(\dfrac{6\,ft.}{2}\right)^3$

Simplify: $\dfrac{4}{3} \cdot \pi \cdot 27$ or $113\,ft.^3$

Multiply the sphere's volume by three: $3 \cdot 113\,ft.^3 = 339\,ft.^3$

And subtract from the cylinder's volume: $770\,ft.^3 - 339\,ft.^3 = 431\,ft.^3$ (approximately)

CHAPTER NINE

Trigonometry

> **Assumptions:** You have sufficient knowledge to take and pass the challenge exam of all previous chapters. You have a scientific calculator and basic knowledge of how to use it.
>
> **Application:** Required for ASP, CSP, OHST, and CHST examinations.
>
> **Discussion:** Understanding trigonometry is essential because many real-world problems require the ability to determine information from triangles. The ability to use trigonometry to calculate lengths and stresses is a skill that will be required both in safety certification examinations and in real life. If you feel that your comprehension of trigonometry is good, go to the challenge exam at the end of this chapter. If you can work all of the problems correctly, skip this chapter.

Trigonometry is the study of triangles and the relationships between their angles and sides. The sum of the internal angles (measured in degrees) of all triangles is 180. This means that whether or not a triangle is a right triangle (one with one internal angle equal to 90 degrees), the sum of all internal angles is 180. This characteristic is not given in the *Candidates' Handbook*, is relatively simple, and you are expected to know it.

$45 + 45 + 90 = 180$ $30 + 60 + 90 = 180$ $60 + 60 + 60 = 180$

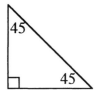 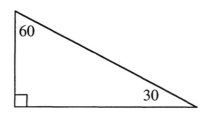 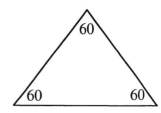

145

Special Properties of Right Triangles

- A right triangle has one angle of exactly 90 degrees.
- The right angle is designated by a small square in the 90° angle.
- The right angle is arbitrarily labeled "C."
- The side opposite the right angle is the longest side of the triangle, and is called the hypotenuse.
- All sides of the triangle are labeled with the lower case letter of the angle opposite them.

Table 14: Special Properties of Right Triangles

The following diagram is given in the *Candidates' Handbook*:

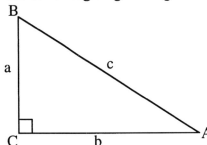

All trig material provided from here on uses these labeling conventions.

Trigonometric Functions

The Sine, Cosine, and Tangent functions are provided in the *Candidates' Handbook*. There are additional trig functions (cotangent, secant, cosecant) but they fall beyond the scope of this book and are not typically required for Safety Certification examinations. Functions that are given (and will be needed) include the following, which are based on the diagram of a standard right triangle shown earlier:

Sine	such that: $\sin A = \dfrac{a}{c}$ (or the "opposite over the hypotenuse")
Cosine	such that: $\cos A = \dfrac{b}{c}$ (or the "adjacent over the hypotenuse")
Tangent	such that: $\tan A = \dfrac{a}{b}$ (or the "opposite over the adjacent")

Table 15: Trig Functions

To understand this clearly, use the following example:

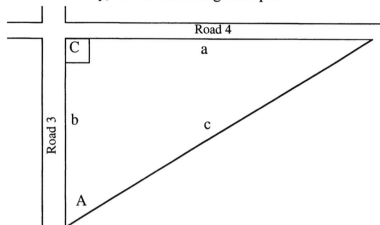

A triangular lot sits at the intersection of Roads 3 and 4. The diagram above is an aerial view of the property. Because the roads intersect at a right angle, the shape of the property is a right triangle. If you stand at the intersection of roads 3 and 4, looking out over the property, the far side (side *c*) is the side *opposite* the intersection. Side *c* is considered to be "opposite" angle C. The side across from the right angle is called the *hypotenuse*.

Because of the characteristics of right triangles, the hypotenuse is always the longest side of a right triangle. It is now established that the angle at the intersection of roads 3 and 4 is a right (or 90-degree) angle, that side *c* (the hypotenuse) is the longest side of the property, and that the right angle of any triangle is always (arbitrarily) labeled C.

Now walk down Road 3 to the corner of the property marked "A" on the diagram. Standing in corner A, look out over the property. The side across from angle A-the side along Road 4, or side *a*-is considered to be "opposite" angle A. The remaining side (the side along Road 3, or side *b*) is described as *adjacent* to angle A. For either of the non-right angles of a right triangle, the "adjacent" side always touches the corner of that angle, and is not the hypotenuse.

Once you understand the relationships of the sides of a right triangle to angle "A" (opposite, adjacent, and hypotenuse), trig functions can provide a wealth of information about the triangle.

If you know that side *b* (along Road 3) is three miles long, and side *a* (along Road 4) is four miles long, the functions given in the *Candidates' Handbook* can help you find other information about the triangle:

EXAMPLE

Find the angle A in degrees. Because $a = 4$ and $b = 3$, and a and b are opposite and adjacent, respectively, to angle A, select a trig function based on this information. The function that uses the lengths of sides a and b (the opposite and the adjacent) is the tangent: $\tan A = \frac{a}{b}$. To find the measure of angle A, substitute the lengths of a and b: $\tan A = \frac{4}{3}$ or $\tan A = 1.33$ (approximately). On scientific calculators you can enter the tangent of an angle, press a key (usually labeled TAN^{-1}) or a key-sequence (often INV + TAN), and the calculator will produce the angle in degrees. In this case, entering a tangent of 1.33 produces an angle of approximately 53 degrees.

Trig functions can also be used to find the lengths of the sides of a triangle. If you need a fence for side c, you must know how much fencing to purchase. The trig formula for determining the length of sides is the Law of Cosines: $c^2 = a^2 + b^2 - 2ab \cos C$. Use the lengths of sides a and b to determine the length of side c:

Enter the lengths of sides a & b: $\qquad c^2 = 4^2 + 3^2 - (2 \bullet 3 \bullet 4 \bullet \cos C)$

Enter the value of the cosine of C. For a right triangle, C is always the 90-degree angle, and the cosine of 90-degrees is zero. (check this with a calculator if you wish to verify).

This makes the equation: $\qquad c^2 = 16 + 9 - (2 \bullet 3 \bullet 4 \bullet 0)$

Which simplifies to: $\qquad c^2 = 25 - 0$ or $c^2 = 25$

Take the square root of both sides: $\quad \sqrt{c^2} = \sqrt{25}$ which becomes $c = \pm 5$

Since negative-five miles makes no sense in this problem, the length of side c is five miles (which is the length of fencing needed).

This is all fine for angle A along Road 3. What happens if you want to find out information about the angle on Road 4 (not the intersection)? The simplest thing to do is to relabel the diagram:

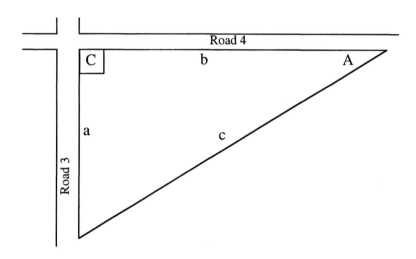

The only rules you must remember for relabeling are these:

Side "c" is always opposite the right angle ("C").

Side "a" is always opposite angle "A."

Note that once the diagram is relabeled, side a is three miles long, and side b is four. Side c remains 5 miles long.

EXAMPLE

If you only know that side b was four miles long and that side c was five, how can you find the measure of angle A?

Find the trig function that uses sides b and c—the cosine: $\cos A = \dfrac{b}{c}$. Enter the numeric values: $\cos A = \dfrac{4}{5}$ or $\cos A = 0.8$. Use a calculator to find that the angle with a cosine of 0.8 is approximately 37 degrees. Check this by using the "180 rule:" The sum of the internal angles of a triangle is 180 degrees. Angle C is 90 degrees, so if the angle along Road 4 is 37 and the angle along Road 3 is 53, then it should be true that 90 + 53 + 37 must equal 180; and it is.

Law of Cosines

$c^2 = a^2 + b^2 - 2ab\cos C$ is called the Law of Cosines, and is given in the *Candidates' Handbook*. Showing the implied multiplication symbols, this is the same as: $c^2 = a^2 + b^2 - (2 \bullet a \bullet b \bullet \cos C)$. This formula works with any triangle, whether or not it is a right triangle. Because the cosine of C is always zero for a right triangle, the formula simplifies to $c^2 = a^2 + b^2$ (The Pythagorean Theorem) for right triangles. If you know the lengths of any two sides of a right triangle, the Law of Cosines allows you to calculate the length of the third side. For non-right triangles, you must know the measure of an angle, in addition to the lengths of two sides.

Law of Sines

$\dfrac{a}{\sin A} = \dfrac{b}{\sin B} = \dfrac{c}{\sin C}$ is the Law of Sines, a formula given in the *Candidates' Handbook*.

It says:

The lengths of the sides of a triangle are directly proportional to the sines of the angles opposite them.

The Law of Sines (like the Law of Cosines) is applicable to all triangles, not just right triangles. It is useful in two circumstances:
    If you know any two angles and the side opposite either of them, you can determine the side opposite the second known angle.

EXAMPLE

If angle $A = 30°, B = 45°$, and side $a = 8$ inches, then the law of sines says that the ratio of the length of a (8 inches) to the sine of 30 is equal to the ratio of the length of side b to the sine of 45.

Express mathematically:	$\dfrac{8}{\sin 30} = \dfrac{b}{\sin 45}$
Cross multiply:	$8 \bullet 0.707 = 0.5b$
Perform multiplication:	$5.657 = 0.5b$
Divide both sides by 0.5:	11.3 inches = side b (approximately)

In the second circumstance, if you know the length of any two sides and the angle opposite one of them, you can determine the angle opposite the other known side.

EXAMPLE

If side $a = 20$ inches, side $b = 10$ inches, and angle $A = 45°$, the Law of Sines says that "the ratio of 20 to the sine of 45 equals the ratio of 10 to the sine of angle B."

Express mathematically:	$\dfrac{20}{\sin 45} = \dfrac{10}{\sin B}$
Cross multiply:	$20 \bullet \sin B = 10 \bullet 0.707$
Perform multiplication:	$20 \sin B = 7.07$
Divide both sides by 20:	$\sin B = 0.354$ (approximately)
Use a calculator to find B:	angle B = 21 degrees (approximately)

Caveat: This may not be the correct answer. There are always two angles between 0 and 180° with the same sine. The second angle is the "supplement" (180 minus the angle of interest) to the first. In this case, the second angle is 159 degrees (180 – 21). This situation is impossible to resolve using the Law of Sines. Knowing two sides and the angle opposite one of them is not always enough to find the other angles.

Trigonometry works for forces as well as distances.

Trigonometry can resolve the lengths of the sides and measures of angles using functions of sine, cosine, and tangent. What makes trigonometry so useful to engineers, architects, and safety professionals is that the trig functions also describe the proportions of forces in triangular shapes.

CHAPTER 9

Lengths of the sides of triangles are directly proportional to the forces involved.

EXAMPLE

In the illustration at right, a pole is anchored horizontally in a vertical wall and extends six feet out from the wall. A wire is attached to the top of the wall (four feet above the pole), and extends over the tip of the pole. A 500 pound weight is suspended from the end of the wire. Find the stress applied to the pole, and the stress on the wire between the top of the wall and the end of the pole. Use trigonometry to find these quantities. First, simplify the drawing and label it as a right triangle:

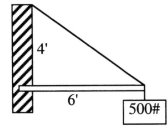

Because the stresses on the sides of the triangle are directly proportional to the lengths of the sides, first find the length of side c. Use the Law of Cosines, given in the *Candidates' Handbook*:

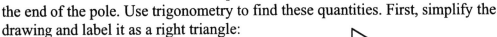

The formula is given as: $c^2 = a^2 + b^2 - 2ab\cos C$

Enter the data:

$$c^2 = 4^2 + 6^2 - (2 \bullet 4 \bullet 6 \bullet \cos C)$$

Since C is a right angle, the cosine of C is zero: $c^2 = 16 + 36 - 0$ or $c^2 = 52$

Take the square root of each side: $\sqrt{c^2} = \sqrt{52}$ yields $c = \pm 7.2$ (approximately)

Since negative 7.2 makes no sense as a length, side c is 7.2 feet long.

Now evaluate stresses. The study of the branch of physics known as "statics" reveals that unless objects are in motion, the stresses acting on them must be balanced by equal and opposite forces. These forces, which have both direction and magnitude, are called "vectors."

Vectors are shown as pointed arrows, with their length representing their magnitude, and their points representing their direction. For the problem, one force involved is the force of gravity. Gravity exerts a force of 500 pounds directly downward on the wire rope that suspends the weight. Because the pole and wire arrangement is in stasis (not in motion), a balancing force of 500 pounds must exist in the opposite direction. Having drawn this "equal and opposite" vector, we can create a "shadow triangle" as shown that is identical in dimensions to the original. Note that the hypotenuse (c) and the "shadow" vector for it (c') share the same line.

Because the length of a triangle's sides are directly proportional to the forces on those sides, set up an equation that says: "The ratio of the length of side a to the 500 pounds of force on a' equals the ratio of the length of side c to the force on c'":

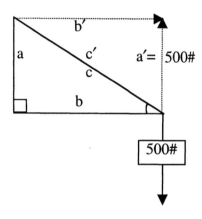

Express mathematically: $\dfrac{length\ a}{force\ a'} = \dfrac{length\ c}{length\ c'}$ or $\dfrac{4\ ft.}{500\#} = \dfrac{7.2\ ft.}{c'\#}$

Cross-multiply: $4\ ft. \bullet c'\# = 7.2\ ft. \bullet 500\#$

Divide both sides by four feet: $c'\# = \dfrac{7.2\ ft. \bullet 500\#}{4\ ft.}$

The "feet" unit cancels out, leaving: $c'\# = \dfrac{7.2 \bullet 500\#}{4}$

Perform multiplication & division: $c'\# = \dfrac{3,600\#}{4}$ or $c'\# = 900\#$

Since the hypotenuse (c in this diagram) represents the wire from the top of the wall to the tip of the pole, and because the hypotenuse is the longest side of the triangle, the force on c' (900 pounds) is the largest force in this arrangement.

The other quantity requested in this problem is the stress on the pole itself. Again, the ratio of the length of side a to the 500 pounds of force on a' equals the ratio of the length of the pole (b) to the force on b'.

CHAPTER 9

Express mathematically: $\dfrac{4\,ft.}{500\#} = \dfrac{6\,ft.}{b'\#}$

Cross-multiply: $4b' = 6 \bullet 500 = 3{,}000$

Divide both sides by 4: $b' = 750$ pounds

The pole is compressed against the building with a force of 750 pounds. It can also be said that the pole exerts an outward force of 750 pounds to resist the compression. An explanation of physics is beyond the scope of this book, but you should understand the directly proportional relationship between the lengths of the sides of a triangle and the stresses on those sides.

Scaffold Problems

Scaffold problems often use proportions of length and stress. A common question might be phrased as follows:

EXAMPLE

A ten-foot tall pylon is a foot in diameter. A line will be hooked to the top of the pylon and side stress applied. If the pylon weighs 500 pounds, how much stress can be horizontally applied to the line at the top before the pylon tips?

The problem is best represented in two different drawings:

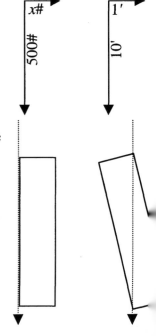

The ratio of the force required to tip the pylon, x, to the weight of the pylon (500 pounds) equals the ratio of the distance required to tip the pylon (one foot) to the distance of that force from the ground (10 feet).

Why is one foot the distance required to tip the pylon? The drawings at right illustrate the reason:

The dashed line represents the point where the base of the pylon is resting on the ground. If the top of the pylon moves one foot horizontally (the diameter of the base), the pylon is exactly balanced. Any additional force tips the pylon.

TRIGONOMETRY

Express the proportions mathematically: $\dfrac{x\#}{500\#} = \dfrac{1\,ft.}{10\,ft.}$

Cross multiply: $10\,ft. \bullet x\# = 500\,ft.\#$

Divide both sides by 10 ft. $x\# = 50\#$

Therefore, if any force more than 50 pounds is applied horizontally to the top of the pylon, the pylon will tip. Note that although this problem initially appeared to involve triangles (and thus, trig functions), the problem was actually solved through simple proportions.

A variant of scaffold problems is commonly found in the field and on safety certification examinations. In this variant, a downward force is applied outside the perimeter of the scaffolding.

EXAMPLE

A 200 pound worker has a 500 pound piece of equipment on a ledge three feet outside the structure of a 10' by 10' by 10' scaffold. The scaffold structure weighs 1,500 pounds. Will the scaffold tip? Draw this as follows:

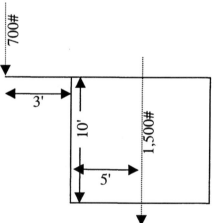

Note that the center of gravity of the scaffold is at the center of the structure (five feet from the pivot point).

The *Candidates' Handbook* gives the formula for such situations as:

$F_1 d_1 = F_2 d_2$, where the F values represent the forces involved, and the d values represent the distances.

Enter the data: $700\# \bullet 3\,ft. = 1{,}500\# \bullet 5\,ft.$

Multiply: $2{,}100\text{ ft.lb.} = 7{,}500\text{ ft.lb.}$

Because the scaffold-side forces are greater, the scaffold will not tip under the circumstances described. Again, notice that although this problem appeared to involve triangles, no trig was required to solve the problem, and knowing the 10-foot height of the scaffold structure was not necessary to solve the problem.

155

CHAPTER 9

Sling Problems

A common use of applied trigonometry is determination of actual stresses on slings when loads are lifted. If all sling loads are vertical, only the only stress is that of the actual weight lifted. This the reason that spreader bars are commonly used to lift loads. The illustrations below show several arrangements that all have stresses of 1,000 pounds per sling.

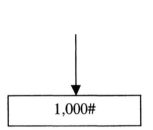 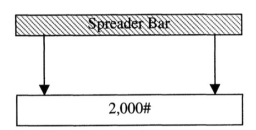 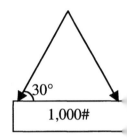

When the sling legs are at an angle to the load, greater stress is imposed on each sling leg. In fact, the more acute the angle of the sling leg to the load, the greater the stress. This relationship is described by the laws of trigonometry.

EXAMPLE

In the drawing above where the two sling legs each make angles of thirty degrees to the plane of the load, the load can be visualized as if there were two vertical sling legs in the center of the load, each supporting half the weight. This is shown at right:

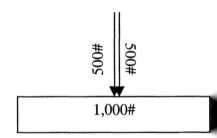

An actual sling is shown at right, that makes a thirty degree angle to the plane of the load. (This angle is selected arbitrarily; slings can be at any angle.) Since the slings (and the load) are symmetrical, it is only necessary to calculate the stress on one side. The stresses that affect one side will be identical for the other.

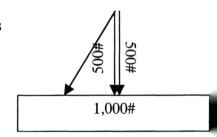

In the diagram at right, the length of the hypotenuse represents the stress on one side of the sling.

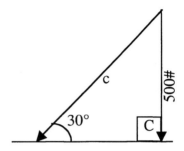

Use the sine function to determine
the stress on the sling: $Sin30 = \dfrac{500}{c}$

Multiply both sides by c: $c \bullet Sin30 = 500$

Divide both sides by the sine of 30: $c = \dfrac{500}{Sin30}$

Enter the data: $c = \dfrac{500}{0.5}$

Perform division: $c = 1,000$

Since the original vector was expressed in pounds, the stress on the sling is 1,000 pounds for each of the two symmetrical legs of the sling.

Be careful on sling problems. The data given by the problem may not always be the angle of the sling to the load, so solve first for the angle of the sling to the load, and then apply the formula.

CHAPTER 9

TRIGONOMETRY QUESTIONS

(NOTE: Triangles are not drawn to scale)

a. Two of the angles of a triangle are 80 and 30 degrees. What is the measure of the remaining angle?

b. Given the triangle at the right, determine the length of side c.

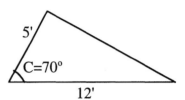

c. Given the triangle at the right, determine angle A.

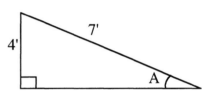

d. Given the triangle at the right, determine angle A.

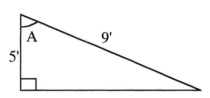

e. Given the triangle at the right, determine angle A.

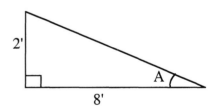

f. Given the triangle at the right, determine the length of the side opposite angle B.

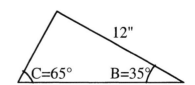

TRIGONOMETRY

g. A two-legged sling is used to lift a two-ton load. The sling straps make an angle of 50 degrees to the horizontal of the load. What is the stress on each strap of the sling?

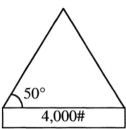

h. A 700-pound scaffold measures 6 feet by 6 feet by 18 feet tall. A six-foot tall man who weighs 200 pounds stands at one edge of the scaffold and pushes against a wrench at head height. What force must the man exert to tip the scaffold?

TRIGONOMETRY ANSWERS

a. 70 degrees, because the sum of all internal angles of a triangle must be 180 degrees:

$$80 + 30 + x = 180 \text{ or } 110 + x = 180$$

Subtracting 110 from both sides: $x = 70$ degrees

b. Approximately 11.31 feet, because the Law of Cosines states:
$$c^2 = a^2 + b^2 - 2ab\cos C$$

Enter the data: $c^2 = 5^2 + 12^2 - (2 \bullet 5 \bullet 12 \bullet \cos 70°)$

Perform multiplication: $c^2 = 25 + 144 - 41.042$

Combine terms: $c^2 = 127.958$

Take the square root of both sides: $c = \pm 11.312$ feet

Since negative 11.31 makes no sense as the length of a side, the correct answer is 11.31 feet.

c. Approximately 35.85 degrees, because the sine function is: $\sin A = \dfrac{a}{c}$

Enter the data: $\sin A = \dfrac{4}{7}$ or $\sin A = 0.571$

Calculate angle A: $A = 34.85$ degrees (approximately)

CHAPTER 9

d. Approximately 56.25 degrees.

    The cosine function is: $\cos A = \dfrac{b}{c}$

    Enter the data: $\cos A = \dfrac{5}{9}$ or $\cos A = 0.556$

    Calculate angle A: A = 56.25 degrees (approximately)

e. Approximately 14.036 degrees

    The tangent function is: $\tan A = \dfrac{a}{b}$

    Enter the data: $\tan A = \dfrac{2}{8}$ or $\tan A = 0.25$

    Calculate angle A: A = 14.036 degrees (approximately)

f. Approximately 7.59 inches

    The Law of Sines states: $\dfrac{b}{\sin B} = \dfrac{c}{\sin C}$

    Enter the data: $\dfrac{b}{\sin 35} = \dfrac{12''}{\sin 65}$ or $\dfrac{b}{0.574} = \dfrac{12''}{0.906}$

    Cross-multiply: $0.906 b = 6.883$

    Divide both sides by 0.906:   b = 7.59 inches (approximately)

g. Approximately 2,611 pounds.

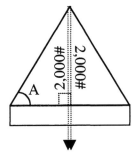

The vector of gravity for the load points directly downward from the center of the hook. Exactly half of the load is carried by side *a* of the right triangle on either side. Since angle A is equal to 50 degrees and the problem asks for the stress on side *c* (the side opposite the right angle formed by the vector of gravity and the plane of the load), use the sine function ($\sin A = \frac{a}{c}$) to determine the stress on side c.

Enter the data: $\quad\quad\quad\quad \sin 50° = \frac{2{,}000\#}{c\#}$ or $0.766 = \frac{2{,}000\#}{c\#}$

Multiply both sides by c: $\quad\quad 0.766c = 2{,}000$ pounds

Divide both sides by 0.766: $\quad\quad c = 2{,}611$ pounds (approximately)

h. Approximately 225 pounds.

First, draw the scaffold and force diagrams:
The ratio of the height of the combined man/scaffold structure to the horizontal distance the pair must move to tip equals the ratio of the combined man/scaffold weight to the horizontal force needed to tip.

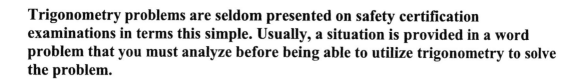

Express mathematically: $\quad\quad \dfrac{24\ ft.}{6\ ft.} = \dfrac{900\#}{x\#}$

Cross-multiply: $\quad\quad\quad\quad\quad 24x = 5{,}400$

Divide both sides by 24: $\quad\quad x = 225$ pounds

Trigonometry problems are seldom presented on safety certification examinations in terms this simple. Usually, a situation is provided in a word problem that you must analyze before being able to utilize trigonometry to solve the problem.

Friction

Before the stress of moving objects on inclined surfaces can be discussed, attention needs to be given to how friction affects the movement of objects on flat surfaces. All objects are subject to a "normal force" represented by the letter N. A "vector" is a force that has both magnitude and direction. The vector of normal (N) force is always at right angles (perpendicular) to the plane on which the object rests. For objects on flat surfaces, the normal force is equal to the force of gravity, and the vector is

CHAPTER 9

commonly expressed as the weight of the object. For example, if a box weighing five pounds sits on a horizontal floor, the normal force (N) is five pounds. In order to slide the box along the floor, friction must be overcome. If both the box and the floor are smooth, very little friction exists. If the box is rough and the floor is carpeted, more friction must be overcome to move the box.

The measure of friction is called the "coefficient of friction," and is represented by the Greek letter *mu* (μ). The lower the mu, the less friction there is.

The force required to move the box across the floor is expressed by the letter "F." The *Candidates' Handbook* gives the entire relationship in the formula: $F = \mu N$. In the five-pound box example, if the mu is one, then it takes five pounds of horizontal force to move the box along the floor. If the mu were 0.2, it would only take one pound of horizontal force to move the box. If the mu were 5, it would take 25 pounds of horizontal force to move the box. These examples are shown below:

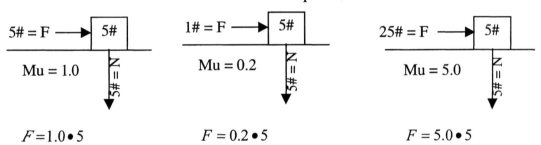

Ramp Problems

EXAMPLE

A cart is used to haul ore from a mine. The total weight of the cart and load is 4,000 pounds. The ramp forms a 25 degree angle with the floor of the mine. The coefficient of friction between the cart wheels and the ramp surface is 0.1. A tagline is tied to the cart and attached to a winch that pulls on the tagline. What amount of force is required to move the cart up the ramp?"

The first step is to define some vectors. The vector of gravity always points vertically downward. The vector of gravity will have a value of 4,000 pounds as given in the problem,

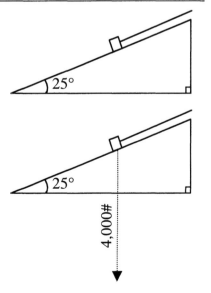

since that is the weight of the cart and its contents.

The second is perpendicular to the ramp surface. This vector can be called the vector of *stick*. This is the "normal" force that, when multiplied by the coefficient of friction, will define the cart's resistance to movement. Note that this resistance applies both to movements up and down the ramp. Also notice that the vector of stick is always at a right angle to the surface that supports the load.

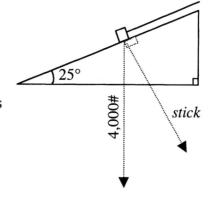

Finally, a third vector is parallel to the ramp surface. This can be called the vector of *slide*. This is the force that would cause the cart to roll down the ramp (ignoring *stick*) if there were no friction at all.

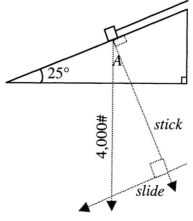

The "vector triangle" is identical to the original in internal angle measurements. Proof of this is beyond the scope of this book, but common geometry texts have additional information if you are interested. Angle A in the vector triangle will always have the same measure as that of the ramp to the horizontal (25 degrees, in this example).

There is now enough data to calculate *stick* and *slide*. If angle A equals 25 degrees, find the opposite side (the vector of *slide*, or side *a*) using the sine

function: $$\sin A = \frac{a}{c}$$

Enter the data: $$\sin 25° = \frac{slide}{4,000}, \text{ or } 0.423 = \frac{slide}{4,000}$$

Multiply both sides by 4,000:   1,690 pounds = *slide* (approximately)

CHAPTER 9

This means that if there were no friction whatsoever, the cart would want to roll down the ramp with a force of about 1,690 pounds.

Now use the cosine function to determine
the vector of *stick*, or side *b*: $\qquad \cos A = \dfrac{b}{c}$

Enter the data: $\qquad \cos 25° A = \dfrac{stick}{4,000}$, or $0.906 = \dfrac{stick}{4,000}$

Multiply both sides by 4,000: $\qquad 3,625$ pounds $= stick$

This means that if the coeffecient of friction were one, the cart would remain where it was on the ramp with a force of 3,625 pounds. The *Candidates' Handbook*, however, provides the formula: $F = \mu N$, where the "F" represents the actual force that will be required, μ (mu) represents the coefficient of friction (given in the problem as 0.1), and the "N" represents the raw value of the vector at right angles to the surface (or the "normal" force). In this problem, the "N" is equal to the value of the vector of *stick* (side *b*).

Substitute values into the formula: $\qquad F = 0.1 \bullet 3,625$ pounds

Perform multiplication: $\qquad F = 362$ pounds (approximately)

Summary: The cart would like to roll down the ramp with a (*slide*) force of 1,690 pounds. A force of friction equal to 362 pounds must be overcome to move the cart in either direction. Therefore, the force required to move the cart up the ramp is that required to overcome friction plus that required to overcome the tendency to slide back down the ramp.

Enter the data: 1,690 pounds + 362 pounds = 2,052 pounds

Thus, any force over 2,052 pounds on the tag line will move the cart up the ramp.

Multiple Triangles

EXAMPLE

A 20-foot ladder is leaned against a wall. The top of the ladder touches the wall at a height of 16 feet. The ladder is moved, and the top of the ladder now touches the wall at a height two feet lower than before. How far did the base of the ladder move?"

To solve this problem, draw "before" and "after" triangles and compare their dimensions.

Draw the "before" triangle:

The Law of Cosines is the best formula to use to find the length of the third side of the triangle.

$$c^2 = a^2 + b^2 - 2ab\cos C$$

Enter data: $20^2 = 16^2 + b^2 - (2 \bullet 16 \bullet b \bullet 0)$

Perform multiplication: $400 = 256 + b^2 - 0$

Subtract 256 from both sides: $144 = b^2$

Take the square root of both sides: $\pm 12 = b$

Since negative 12 feet makes no sense as a distance, the base of the ladder is 12 feet from the building's base in the "before" scenario.

Next, draw the "after" triangle. After the ladder was moved, the top of the ladder touched the building at a height two feet lower than before:

Again use the Law of Cosines: $c^2 = a^2 + b^2 - 2ab\cos C$

Enter data: $20^2 = 14^2 + b^2 - (2 \bullet 14 \bullet b \bullet 0)$

Perform multiplication: $400 = 196 + b^2 - 0$

Subtract 196 from both sides: $204 = b^2$

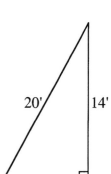

Take the square root of both sides: $\pm 14.3 = b$
(approximately)

Since negative 14.3 feet makes no sense as a distance, the base of the ladder is 14.3 feet from the building's base in the "after" scenario.

Finally, the "before" and "after" distances must be compared. The question asks "How far did the base of the ladder move?" between the two conditions. Subtract the "before" (12 feet) from the "after" (14.3 feet), to find that the base of the ladder moved 2.3 feet (approximately).

TRIGONOMETRY CHALLENGE EXAM

(NOTE: Triangles are not drawn to scale)

a. Two angles of a triangle are 25 and 75 degrees. What is the measure of the third angle?

b. Is the triangle at right a conventionally labeled triangle?

c. In the triangle shown at right, what is the length of side a?

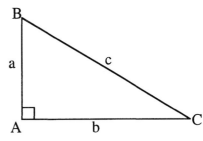

d. In the triangle shown at right, what is the length of side a?

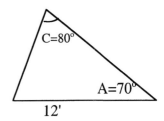

TRIGONOMETRY

e. In the right triangle shown at right, what is the measure of angle A?

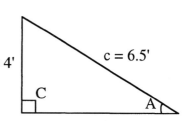

f. In the right triangle shown at right, what is the length of side *c*?

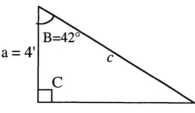

g. In the right triangle shown at right, what is the measure of angle A?

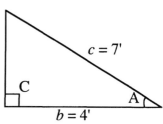

h. In the following right triangle, what is the length of side *b*?

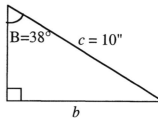

i. In the right triangle at right, what is the measure of angle A?

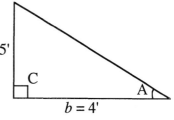

CHAPTER 9

j. In the right triangle at right, what is the length of side *b*?

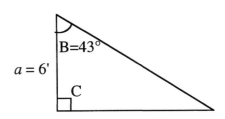

k. Given the following triangle measurements: *a* = 4, *b* = 3, and *c* = 5, what stress exists on side *a* if side *b* is under a stress of 500 pounds?

l. A load is being lifted by a pair of sling straps. The tension on each leg of the sling is 4,000 pounds. Angle theta (θ) is 80 degrees. How much does the load weigh?

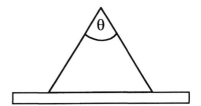

m. A forklift is sitting on a 20 percent grade. The forklift weighs 3,200 pounds, and the brake is set. Conditions are icy, and the coefficient of friction is 0.08. Will the forklift remain in place, or will it slide down the grade, causing potential injury?

n. A 20-foot-long guy wire supports the tip of a pole that is anchored in the masonry of a building wall. The wire is attached to the wall twelve feet above the pole. Changes in the support structure require that the upper attachment point for the guy wire be moved three feet down. How much shorter must the new guy wire be?

TRIGONOMETRY CHALLENGE EXAM ANSWERS

a. 80 degrees, because the internal angles of a triangle is 180 degrees. 25 + 75 + 80 = 180.

b. No, because in a right triangle, the right angle is called C. Also the sides are given lower case letters corresponding to the uppercase letters assigned to the angles opposite them.

c. Approximately 9.6 feet.

   The Law of Cosines states: $\quad c^2 = a^2 + b^2 - 2ab\cos C$

TRIGONOMETRY

Enter the data: $12^2 = a^2 + 9^2 - (2 \bullet a \bullet 9 \bullet \cos 80)$

Perform multiplication: $144 = a^2 + 81 - 3.126a$

Subtract 144 from both sides: $0 = a^2 - 3.126a - 63$

Use the quadratic equation: $x = \dfrac{-b \pm \sqrt{b^2 - 4ac}}{2a}$

Enter the data:
$$a = \dfrac{3.126 \pm \sqrt{(-3.126)^2 - (4 \bullet 1 \bullet (-63))}}{2 \bullet 1}$$

Simplify:
$$a = \dfrac{3.126 \pm \sqrt{9.770 + 252}}{2}$$

Simplify again:
$$a = \dfrac{3.126 \pm \sqrt{261.770}}{2}$$

Simplify yet again: $a = \dfrac{3.126 \pm 16.179}{2}$ which yields two equations:

First:
$$a = \dfrac{3.126 + 16.179}{2}$$

Perform addition: $a = \dfrac{19.305}{2} = 9.652$

Second: $a = \dfrac{3.126 - 16.179}{2}$

Perform subtraction: $a = \dfrac{-13.053}{2} = -6.526$

Since negative 6.526 makes no sense as a side length, the correct answer is approximately 9.6 feet.

d. Approximately 11.4 feet.

The Law of Sines states:	$\dfrac{a}{\sin A} = \dfrac{b}{\sin B} = \dfrac{c}{\sin C}$	
Enter the data:	$\dfrac{a}{\sin 70} = \dfrac{12'}{\sin 80}$	
Use a calculator to find sines:	$\dfrac{a}{0.940} = \dfrac{12'}{0.985}$	
Cross-multiply:	$0.985a = 0.940 \bullet 12$	
Simplify:	$0.985a = 11.276$	
Divide both sides by 0.985:	$a = 11.4$ feet approximately	

e. Approximately 38 degrees.

The sine function is:	$\sin A = \dfrac{a}{c}$
Enter the data:	$\sin A = \dfrac{4}{6.5}$ or $\sin A = 0.615$
Use a calculator to find A:	$A = 38°$ (Approximately)

f. Approximately 5.4 feet.

First, find angle A:	$A = 180 - (90 + 42)$ or $A = 48$ degrees
The sine function is:	$\sin A = \dfrac{a}{c}$
Enter the data:	$\sin 48 = \dfrac{4}{c}$
Use a calculator to find sin 48:	$0.743 = \dfrac{4}{c}$
Multiply both sides by c:	$0.743c = 4$
Divide both sides by 0.743:	$c = 5.4$ (approximately)

g. Approximately 55.2 degrees.

TRIGONOMETRY

    The cosine function is: $\qquad \cos A = \dfrac{b}{c}$

    Enter the data: $\qquad \cos A = \dfrac{4}{7}$ or $\cos A = 0.571$

    Use a calculator to find A: $\qquad A = 55.2$ degrees

h. Approximately 6.2 inches, because:

    First, find angle A: $\qquad A = 180 - (90 + 38)$ or $A = 52$ degrees

    The cosine function is: $\qquad \cos A = \dfrac{b}{c}$

    Enter the data: $\qquad \cos 52 = \dfrac{b}{10}$

    Use a calculator to determine $\cos 52$: $\qquad 0.616 = \dfrac{b}{10}$

    Multiply both sides by 10: $\qquad 6.2 = b$

i. Approximately 32 degrees.

    The tangent function is: $\qquad \tan A = \dfrac{a}{b}$

    Enter the data: $\qquad \tan A = \dfrac{2.5}{4}$ or $\tan A = 0.625$

    Use a calculator to determine A: $\qquad A = 32$ degrees, approximately

j. Approximately 5.6 feet.

    First, find angle A: $\qquad A = 180 - (90 + 43)$ or $A = 47$ degrees

    The tangent function is: $\qquad \tan A = \dfrac{a}{b}$

    Enter the data: $\qquad \tan 47 = \dfrac{6}{b}$

Use a calculator to find tan 47: $1.072 = \dfrac{6}{b}$

Multiply both sides by b: $1.072b = 6$

Divide both sides by 1.072: $b = 5.6$ approximately

k. Approximately 667 pounds.

The length of the sides is directly proportional to the stress on them.

Therefore: $\dfrac{4}{a} = \dfrac{3}{500\#}$

Cross-multiply: $2{,}000\# = 3a$

Divide both sides by 3: $a = 667$ pounds, approximately

l. Approximately 6,128 pounds.

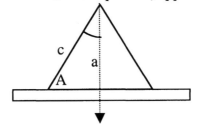

First, draw the vector of gravity (this is the quantity the problem asks for). Because angle theta was 80 degrees, the half of it that applies to the vector triangle is 40 degrees. Since the internal angles of any triangle sum to 180 degrees, subtract 90 (the right angle) and 40 (the angle at the top) from 180 to find angle A = 50 degrees. The hypotenuse value is 4,000 pounds, and the unknown is the side opposite angle A (side a).

The sine function is: $\sin A = \dfrac{a}{c}$

Enter the data: $\sin 50 = \dfrac{a}{4{,}000}$

Note that since each leg of the sling had a stress of 4,000 pounds, the hypotenuse of either side will be 4,000 pounds.

Use a calculator to determine sin 50: $0.766 = \dfrac{a}{4{,}000}$

Multiply both sides by 4,000:   3,064 = a (approximately)

Because the slings are symmetrical, the total weight of the load is 2 times 3,064, or approximately 6,128 pounds.

m. The forklift will slide down the grade.

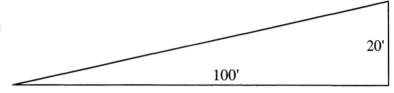

First, the measurements of the ramp must be converted from "percent grade" (as given in the problem) to degrees. The term "percent grade" means that for every hundred units of run, there will be a rise specified by the "percent grade" number. In the case, for every hundred feet of run, the incline rises by 20 feet. This is illustrated in the triangle below:

To find the angle to the left (angle "A"), in degrees, use the tangent function:

$$\tan A = \frac{a}{b}.$$

Enter the data: $\tan A = \frac{20}{100}$ or $\tan A = 0.2$

Use a calculator to determine angle A: A = 11.310 degrees (approximately)

Now construct the vector triangle:

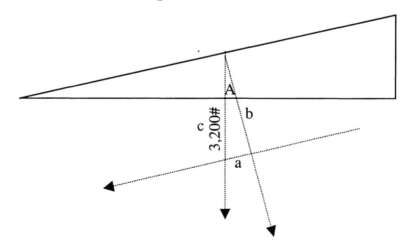

Use the sine function to calculate the vector of *slide*:

$$\sin A = \frac{a}{c}$$

Enter the data:

$$\sin 11.31 = \frac{a}{3{,}200} \text{ or } 0.196 = \frac{a}{3{,}200}$$

Multiply both sides by 3,200: (approximately)

$627.572 = a$ or *slide* $= 628$ pounds

Use the cosine function to calculate the vector of *stick*:

$$\cos A = \frac{b}{c}$$

Enter the data:

$$\cos 11.31 = \frac{b}{3{,}200} \text{ or } 0.980 = \frac{b}{3{,}200}$$

Multiply both sides by 3,200: (approximately)

$3{,}138 = b$ or *stick* $= 3{,}138$ pounds

Now account for the coefficient of friction via the formula:

$$F = \mu N$$

Enter the data:

$F = 0.08 \bullet 3{,}138$ or $F = 251.04$ pounds

Because the forklift wants to slide down the ramp with a force of 628 pounds and has only 251 pounds of friction to keep it in place, it will be sliding down the ramp with a force of 377 pounds (*slide-stick* or 628 pounds minus 251 pounds).

n. Approximately 1.6 feet shorter. Draw the first arrangement:

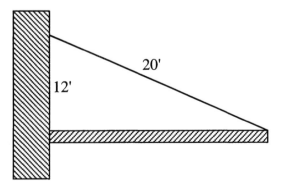

Use the Law of Cosines to find
the length of the pole: $\qquad c^2 = a^2 + b^2 - 2ab\cos C$

Enter the data:
$$20^2 = 12^2 + b^2 - (2 \bullet 12 \bullet b \bullet \cos 90)$$

Simplify:
$$400 = 144 + b^2 - 0$$

Subtract 144 from both sides: $\qquad 256 = b^2$

Take the square root of both sides: $\qquad \pm 16 = b$

Since negative 16 feet makes no sense as a length, the pole is 16 feet long.

Draw the second arrangement:

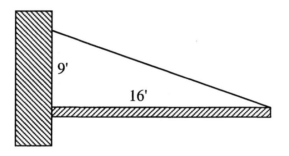

Use the Law of Cosines to find
the length of the new guy wire: $\qquad c^2 = a^2 + b^2 - 2ab\cos C$

Enter the data: $\qquad c^2 = 9^2 + 16^2 - (2 \bullet 9 \bullet 16 \bullet \cos 90)$

Simplify: $\qquad c^2 = 81 + 256 - 0$ or $c^2 = 337$

Take the square root of both sides: $\qquad c = \pm 18.358$ (approximately)

Since negative 18 feet makes no sense as a length, the new guy wire is approximately 18.4 feet long.

Because the question asked the *difference* between the original and new guy-wire lengths, the correct answer is 1.6 feet (20 feet minus 18.4 feet).

CHAPTER 9

BRAIN TEASER

(For those who like trigonometry—no questions of this difficulty are likely to be encountered on safety certification examinatitons):

A ramp makes a 20 degree angle to the horizontal. A box is being pulled up the ramp. 150 pounds of force must be exerted to move the box up the ramp. The coefficient of friction between the box and the ramp is 0.2. How much does the box weigh?

BRAIN TEASER ANSWER

Approximately 283 pounds.

Draw the ramp:

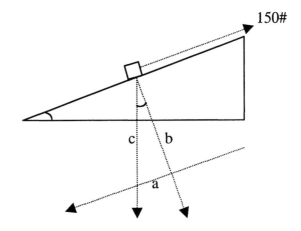

First, define the vectors:

The force required to move the box up the ramp (given as 150 pounds) must equal the vector of *slide* (vector a) plus the vector of *stick* (vector b) times the coefficient of friction.

Express mathematically: $a + (0.2b) = 150$

Subtract a from both sides: $0.2b = 150 - a$

Divide both sides by 0.2: $b = \dfrac{150 - a}{0.2}$

Now solve for a numeric value for vector *a* by using the *b* value calculated above. Do this by using the tangent function for angle A (which is 20 degrees). The tangent was selected because it used both sides *a* and *b*.

The tangent function is: $\tan A = \dfrac{a}{b}$

176

Enter the data:	$\tan 20 = \dfrac{a}{\left(\dfrac{150-a}{0.2}\right)}$
Simplify:	$0.364 = \dfrac{a}{1} \cdot \dfrac{0.2}{150-a}$
Perform multiplication:	$\dfrac{0.364}{1} = \dfrac{0.2a}{150-a}$
Cross-multiply:	$54.6 - 0.364a = 0.2a$
Add $0.364a$ to both sides:	$54.6 = 0.564a$
Divide both sides by 0.564:	$96.8 = a$ (approximately)
Now solve for side c:	
The sine function is:	$\sin A = \dfrac{a}{c}$
Enter the data:	$\sin 20 = \dfrac{96.8}{c}$
Simplify:	$0.342 = \dfrac{96.8}{c}$
Multiply both sides by c:	$0.342c = 96.8$
Divide both sides by 0.342:	$c = 283$ pounds (approximately)

CHAPTER TEN

Boolean Algebra

> **Assumptions:** You have sufficient knowledge to take and pass the challenge exam of all of the algebra chapters. You have a scientific calculator and basic knowledge of how to use it.
>
> **Application:** Required for ASP and CSP examinations.
>
> **Discussion:** Boolean algebra, unlike the type of algebra discussed previously in this book, is the "algebra of events." George Boole developed this system of mathematics in the 1850s. Using basic operators such as AND, OR, and sometimes NOT; Boolean algebra is useful in computer science, digital switching circuits, and Fault Tree analysis.

As used in the *Candidates' Handbook*, the plus symbol (+) represents the word "OR," and the multiplication symbol (•) represents the word "AND." Probabilities must always fall between 0 (an event does not occur) and 1 (the event occurs). Capital letters are commonly used in Boolean algebra to represent events. If "A" represents the situation "event A happens," then A' (pronounced *A prime*) represents the situation "event A does not happen." Using A, B, and C to represent events, the table below interprets the Boolean Postulates provided in the *Candidates' Handbook*:

$A + B = B + A$	The probability that A happens OR B happens is the same as if B happens OR A happens (order doesn't matter).
$A \bullet B = B \bullet A$	The probability that A happens AND B happens is the same as if B happens AND A happens (order doesn't matter).
$A(B \bullet C) = (A \bullet B)C$	The probability that A AND B AND C happen is the same no matter how the events are grouped—the grouping doesn't change the probability.
$A + (B + C) = (A + B) + C$	The probability that A OR B OR C happen is the same no matter how the events are grouped—grouping doesn't change the probability.
$A(B + C) = (A \bullet B) + (A \bullet C)$	The probability that A AND either of B OR C happen is the same as the probability of either A AND B happening OR A AND C happening (A must happen in either case).

$A+(B\bullet C)=(A+B)\bullet(A+C)$	The probability that A will happen OR B AND C will both happen is the same as the probability that A OR B will happen AND A OR C will happen.
$A+A'=1$	The probability that A happens OR A does not happen is absolute—one (1). The event either happens or it doesn't.
$A\bullet A'=0$	The probability that A happens AND A does not happen is impossible—zero (0). The event can't both happen and not happen.
$A\bullet A=A$	The probability that A happens AND A happens is equal to the probability that A happens alone. This does not mean that event A happens twice. It means that the same event may be used different places in the fault tree. Since it is the same event, if A happens anywhere then A happens everywhere in the scenario. (As used in logic-circuit design: If an AND gate is fed inputs of A and A, the output of the gate will be the value of A; 0 if A is false or 1 if A is true.)
$A+A=A$	The probability that A happens OR A happens is equal to the probability that A happens alone. This does not mean that event A happens twice. The same event may be used different places in the fault tree. Since it is the same event, the probability that A happens anywhere in a scenario means that A happens everywhere. (As used in logic-circuit design; if an OR gate is fed inputs of A and A, the output of the gate will be the value of A; 0 if A is false or 1 if A is true.)
$A+0=A$	If an OR gate is fed inputs of 0 and A, the output will be the value of A; 0 if A is false, 1 if A is true.
$A\bullet 1=A$	If an AND gate is fed inputs of 1 and A, the output will be the value of A; 0 if A is false, 1 if A is true.

Table 16: Boolean Postulates

Note that the final two Boolean postulates are most frequently used in logic circuit design. The OR and AND gates used in the final two postulates have input and output values of either 0 (false) or 1 (true). All other Boolean postulates are used either in

Fault Tree analysis (where probabilities are usually some decimal between 0 and 1) or digital circuit design (where values may be 0 and 1 only).

Fault Tree Analysis

For safety certification examinations, Boolean algebra is used most commonly in fault tree scenarios. Fault tree analysis uses deductive logic in a diagram form to examine undesired events. Fault tree analysis places a single, undesired event at the "top event" position. Scenarios that might reasonably lead to the top event are diagramed below it using logic gates and event symbols. Common symbols used in fault tree analysis are listed below:

○	The circle represents a basic component failure.
◇	The diamond represents an event with insufficient significance or information to develop a failure. All analysis must stop at this point until additional information is available.
⌂	This shape represents a normally occurring condition or event with a probability of close to 1.
▭	The rectangle is a combination of any of the above events, and never occurs at the lowest level of the fault tree

Table 17: Common Fault Tree Symbols

Between event symbols, the fault tree has *logic gates*. The most commonly used gates are the AND gate and the OR gate. A hexagonal gate called an INHIBIT gate is also sometimes used when certain conditions or timing are required to allow the scenario to proceed. Additional gates exist (NOT, EQU, NOR, NAND, XOR), but they are beyond the scope of certification exams. The symbols for logic gates are shown below:

CHAPTER 10

⌂	AND Gate: All inputs (bottom) must be true before output (top) is true. Note the flat line at the bottom of the gate.
⌒	OR Gate: Any single input (bottom) must be true before output (top) is true. Note the curved line at the bottom of the gate.
⬡	INHIBIT Gate: This gate is shown as a six-sided figure (hexagon). Think of this as a modified AND gate: Both the input (bottom) AND the side condition (normally shown in a balloon or rectangle connected to the side of the gate) must be true before output (top) is true.

Table 18: Logic Gates

These symbols are not be labeled on certification exams, so you will need to memorize their shapes and functions. A convenient way to remember the difference between the AND and OR gates is to look at the bottom of the symbol: AND gates have a straight line on the bottom of the symbol—like the straight crossbar in the letter "A." OR gates have a curved line on the bottom of the symbol—like the top of the letter "O."

Putting these Boolean concepts into practice with fault tree analysis, the form at right shall serve as a simplified example:

$B = E \bullet F$
$C = G + H - (G \bullet H)$
$D = I + J - (I \bullet J)$
$A = B \bullet C \bullet D$

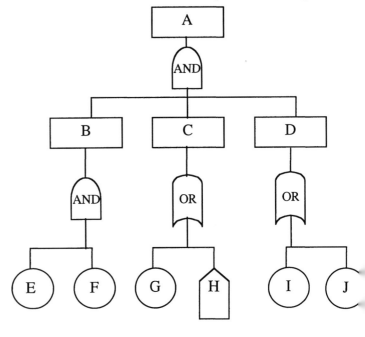

Note that on the OR gate leading to event C, the probability that both G and H could simultaneously fail had to be subtracted from the probability that either failed alone (satisfying the OR gate). The same situation happened on the OR gate leading to event D with components I and J. The reason for this is illustrated in the Venn diagram below:

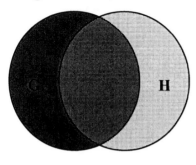

Note the intersection between events G and H. If we merely add G and H (G + H), the intersection will be counted twice, which leads to an unrealistically high probability. The overlap area (the intersection, or $G \bullet H$) must be subtracted to find the true probability of both events happening.

If a probability of 0.25 is arbitrarily assigned to each of components E-G and I-J, the probabilities are as follows:

$B = 0.25 \bullet 0.25$ or 0.0625

$C = 0.25 + 1 - (0.25 \bullet 1)$ or $(1.25 - 0.25)$ or 1

Note that event H is a normally occurring event with a probability at or near one.

$D = (0.25 + 0.25) - (0.25 \bullet 0.25) = 0.4375$

This means, finally, that the probability of the unwanted (top) event is
$A = 0.0625 \bullet 1 \bullet 0.4375$ or $A = 0.2734375$. This rounds to approximately 0.27 or a 27% likelihood that event A would occur.

CHAPTER 10

Another example: consider the following diagram

$B = D \bullet E$
$C = D + F - (D \bullet F)$
$A = [B + C - (B \bullet C)] \bullet 0.5$

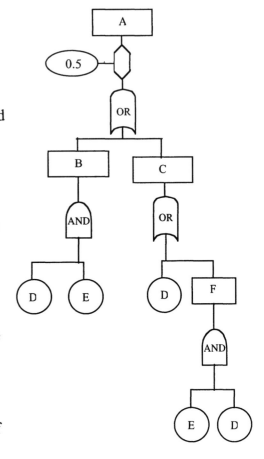

Notice, however, that if component D and the inhibit gate fail, the top event (A) is satisfied with no other events required. The fact that $D \bullet$ inhibit gate alone can cause the top event makes D and the inhibit gate the "least cut set." The fewest possible events that satisfy the requirements of achieving the top event are always considered the least cut set for a given scenario. Obviously, there is no sense in spending resources trying to improve component E since the failure of D and the inhibit gate alone can cause the unwanted event A. In fact, if failure D can be engineered out of the scenario, A becomes impossible.

Always look for the least cut set in any fault tree scenario. The probability of the least cut set is the approximate probability of the top event.

184

BOOLEAN ALGEBRA CHALLENGE EXAM

a. If the probability of event A is 0.65, what is the probability of A'?

b. Event A and either of events (B or C') must happen for a specific situation to occur. Is this probability the same as that of either event (A and B) or (A and C')?

c. The symbol at right is used in fault-tree analysis to represent what?

d. In the fault tree diagram shown below, what is the least cut set?

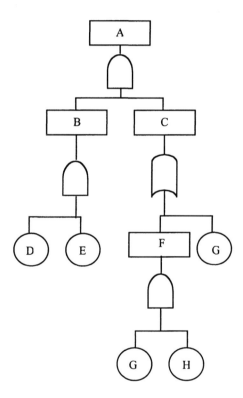

e. Given the following probabilities for the fault-tree diagram shown in question d, what is the probability of the top event? $D = 0.15$, $E = 0.22$, $G = 0.4$, $H = 0.7$

CHAPTER 10

f. Given the fault tree diagram below, with all discrete events and probabilities marked, what is the probability of the top event?

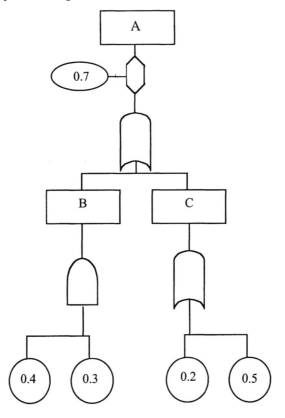

BOOLEAN ALGEBRA CHALLENGE EXAM ANSWERS

a. 0.35 because the postulate (given) says $A + A' = 1$, the probability of A is given as 0.65. Therefore, $0.65 + A' = 1$ becomes $A' = 1 - 0.65$ or $A' = 0.35$. Note that the probability that event A happened (0.65) OR that event A did not happen (A'), must always add up to one.

b. Yes, because the postulate (given) says $A(B+C) = (AB) + (AC)$. If a probability of 0.25 is assumed for event A and a probability of 0.4 for events B and C, the postulate can be proved mathematically:

$$0.25(0.4+0.4) = (0.25 \bullet 0.4) + (0.25 \bullet 0.4)$$

Simplify: $\quad 0.25(0.8) = (0.1) + (0.1)$

Simplify again: $\quad 0.2 = 0.2$

c. A basic component failure.

d. $D \bullet E \bullet G$, because if events D AND E happen, event B occurs. Failure G alone will suffice to cause event C. Once both B and C occur, the top event (A) occurs.

e. Approximately 0.01 (or a 1% chance), because the least cut set ($D \bullet E \bullet G$) is $0.15 \bullet 0.22 \bullet 0.4 = 0.0132$, which rounds to 0.01. Notice that if you got question d wrong (the least cut set) you will miss this one as well. Cascade questions of this type are common on safety certification exams.

f. Approximately 45%, because the probability of event B is $0.4 \bullet 0.3$, or 0.12. The probability of event C is $0.2 + 0.5 - (0.2 \bullet 0.5)$, which simplifies to 0.7-0.1 or 0.6. Note that since the OR gate leading to event C could be satisfied by either of the basic component failures shown, the probability that both components fail is subtracted from the sum of the probabilities. Event A can occur if either event B (0.12) OR event C (0.6) occurs and survives the inhibit gate. Again, consider the possibility that B and C occur simultaneously. The probability of B OR C is $0.12 + 0.6 - (0.12 \bullet 0.6)$, which simplifies to 0.72-0.072 or 0.648. Finally, the inhibit gate must be passed: $0.648 \bullet 0.7 = 0.4536$, which is the final probability of event A.

ABBREVIATIONS AND SYMBOLS USED IN THIS BOOK

ABIH	American Board of Industrial Hygiene
ASP	Associate Safety Professional—a credential offered by the BCSP
BCSP	Board of Certified Safety Professionals—an independent certifying organization
CCHEST	Committee for the Certification of Health, Environmental, and Safety Technologists—a committee associated with the BCSP and American Board of Industrial Hygiene (ABIH)
CHST	Construction Health and Safety Technologist—a credential offered by CCHEST
CIH	Certified Industrial Hygienist—a credential offered by ABIH
CSP	Certified Safety Professional—a credential offered by the BCSP
OHST	Occupational Health and Safety Technologist—a credential offered by CCHEST
$\lvert\ \rvert$	Absolute value symbol
$+$	Addition symbol
$-$	Subtraction symbol or symbol for negative numbers
$\cdot$	Multiplication symbol
$\div$	Division symbol
$=$	Is equal to
$\neq$	Is not equal to
$\geq$	Is greater than or equal to
$\leq$	Is less than or equal to
$\pm$	Numeric value can be positive or negative
atm	One Atmosphere (at sea level) which is equal to 14.7 psia

Table 19: Abbreviations and Symbols Used

INDEX

Absolute Pressure, 53
Absolute Value, 27
Algebraic Properties, 75–77
    Associative, 76
    Commutative, 75
    Distributive, 77
Algebraic Variables, 75
Ampere, 53
AND Gate, 181
Antilogs, 40
Anti-Natural Logs, 40
Area, 131
Associative Property, 76
Atmosphere, 54

Boolean Algebra, 179–84
    Definition, 179
    Fault Tree Analysis, 181
        Common Symbols, 181
        Logic Gates, 181
    Postulates, 179

Calculator
    Additional Functions, 2
    BCSP Rules, 1
    Exam Strategies, 4
    Hierarchy of Operations, 3
    Required Minimum Functions, 1
Candela, 53
Celsius, 49
Centigrade, 49
Circle, 131
Circumference, 131
Coefficient of Friction, 162
Common Denominator, 17
Commutative Property, 75
Cone, 132
Conversions, 55–57
    Dimensional Analysis, 58
    Multipliers, 56
Cosecant, 146
Cosine, 146

Cosines, Law of, 150
Cotangent, 146
Cube, 132
Cubic Feet, 45
Cylinder, 132

Decimals, 5
Diameter, 131
Dimensional Analysis, 58
Direct Proportions, 92
Distributive Property, 77
Dyne, 52

Elements, Set, 103
Engineering Notation, 71–72
    Definition, 71
English Units, 45
Equations, 82–92
    Identity, 82
    Multi Variable, 87
    Operations to Both Sides, 83
    Simultaneous, 88
    Single Variable, 84
Exponents, 31–35
    Definition, 31
    Rules, 32

Factoring, 78
Fahrenheit, 47
Fault Tree Analysis, 181
Feet of Water, 54
Fractions, 5–20
    Adding and Subtracting, 16
        Variables, 17
    Conversions-decimal/percentages, 6
    Decimals, 5
    Definition, 5
    Dividing, 11
        More than one Fraction, 13
    Multiplying, 8
        Whole Numbers by Fractions, 9
    Negative, 16

INDEX

Percentages, 6
Proportions, 22
Reciprocals, 21–22
Rounding, 24–28
    Significant Digits, 24
Friction, 161
    Coefficient of Friction, 162

Gallons, 46
General Form (Quadratic), 112
Geometry, 131–38
    Area, 131
    Circumference, 131
    Common Shapes, 131
    Definition, 131
    Diameter, 131
    Perimeter, 131
    Radius, 131
    Volume, 131
Grams, 48
Graphs, 108–9
    Slope, 108
    Slope Formula, 108

Hierarchy of Operations (Calculator), 3

Inches of Mercury, 54
INHIBIT Gate, 181
Inverse Proportions, 93

Kelvin, 53
Kilopascal, 54

Least Common Denominator, 17
Like Terms, 78
Liters, 49
Logarithms, 39–41
    Antilogs, 40
    Anti-Natural Logs, 40
    Definition, 39
    Natural Logs, 40
Logic Gates, 181

Measurements, 45–57
    Absolute Pressure, 53
    Absolute Temperature, 53
    British Engineering, 52
    English Absolute, 52
    English Units, 45
        Cubic Feet, 45
        Distance, 47
        Gallons, 46
        Pound, 45
        Temperature, 47
    Metric Absolute, 52
    Metric System, 48
        Basis, 51
        Distance, 49
        Grams, 48
        Liters, 49
        Prefixes, 49
        Temperature, 49
        Volume, 48
    SI, 52
    Time, 46
Meters, 49
Meters, Cubic, 48
Metric System, 48
Miles, 47
Mixtures, 106–8
mm Hg, 54
Mole, 53
Mu, 162
Multi-Variable Equations, 87

Natural Logs, 40
Newton, 52
Normal Force, 161

OR Gate, 181
Order, 36

Parallelogram, 131
Percentages, 6
Perimeter, 131
Polynomials, 78
Poundal 54
Pounds, 45
Proportions, 22
    Direct, 92
    Inverse, 93
PSIA, 54
PSIG, 54

INDEX

Quadratic Equations, 112–14
    Formula, 112

Radical, 36
Radicand, 36
Radius, 131
Ramp Problems, 162
Rankin, 53
Reciprocals, 21–22
Rectangle, 131
Rectangular Solids, 132
Roots, 36–39
    Definition, 36
    Expressed as Exponents, 37
    Fractional Exponents, 38
    Higher Orders, 37
    Negative Radicands, 36
    Order, 36
    Radical, 36
    Square Root, 36
Rounding, 24

Scaffold Problems, 154
Scientific Notation, 67–71
    Calculator Keys, 68
    Definition, 67
    Division, 69
    Multiplication, 69
Secant, 146
Sets, 103–6
    Definition, 103
    Elements, 103
    Naming, 103
    Subsets, 103
    Symbols, 104
    Venn Diagrams, 105
Significant Digit, 24
Simultaneous Equations, 88
Sine, 146
Sines, Law of, 150
Single-Variable Equations, 84
Slide, 163

Sling Problems, 156
Slope of a Line, 108
Slug, 52
Sphere, 132
Square, 131
Square Root, 36
Stick, 163
Subsets, 103
Sum of Angles, 145

Tangent, 146
Time, 46
Torr, 54
Trapezoid, 131
Triangle, 131
Trigonometry, 145–66
    Definition, 145
    Forces, 151
    Functions, 146–49
    Law of Cosines, 150
    Law of Sines, 150
    Multiple Triangles, 165
    Properties of Right Triangles, 145
    Ramp Problems, 162
    Relabeling Triangles, 149
    Scaffold Problems, 154
    Sling Problems, 156
    Sum of Angles, 145

Variables, 75
Vector, 161
Venn Diagrams, 105
Volume, 131

Word Problems, 114–20
    Distance Problems, 115
    Math Example, 117
    Method, 114
    Multi Rate/Time, 119
    Ratios and Proportions, 118

Zero Product Property, 112

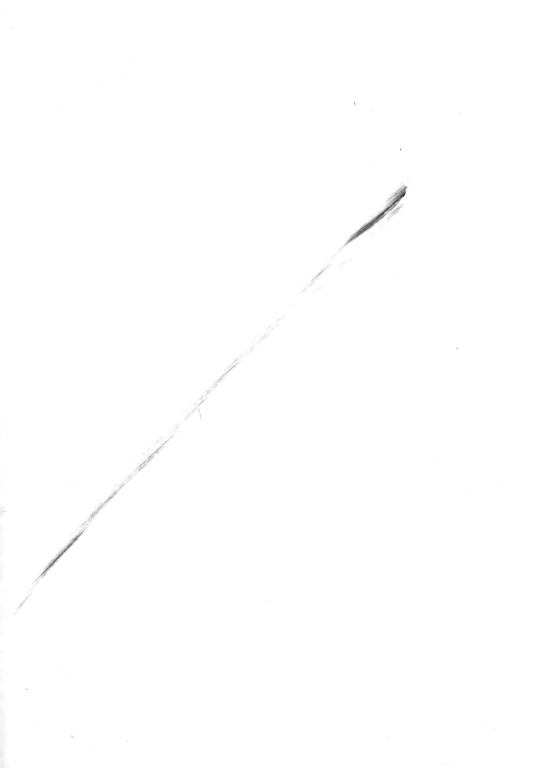